Distribution Network Analysis for Water Utilities

AWWA MANUAL M32

First Edition

FOUNDED 1881

American Water Works Association

American Water Works Association
6666 West Quincy Ave.
Denver, CO 80235

Printed in USA

ISBN 0-89867-457-3

Contents

Preface

The Computer Assisted Design of Water Systems (CADWS) Committee in the Engineering and Construction Division of AWWA was formed in 1982. The scope of the committee is to collect and evaluate for dissemination information concerning the use of computer technology in the design, analysis, and operation of water systems.

The CADWS Committee originally had two tasks. The first task was to conduct and compile the results of a network-analysis questionnaire. The results of the questionnaire were published in January 1984 and are discussed in chapter 1 of this manual. The second task was to prepare this manual.

The purpose of this manual is to compile, discuss, and explain matters relating to the computerized analysis of water-distribution-system networks for use by engineers, planners, managers, and other people involved in the design and operation of water systems. This manual can serve as a reference guide and as a manual of practice for those involved in making decisions regarding the implementation and use of computerized network-analysis programs. The glossary at the end of the book is a guide to the terms used in network analysis.

This manual covers a variety of topics that are of primary concern to people involved in water-system analysis of closed-conduit networks or treated-water systems. Major topics include model preparation, distribution-system network analysis, optimization, and graphics-support software.

Chapter 1 is an introduction to network analysis and computer network-modeling tools. A historical overview is provided to give the reader a sense of the evolution of computerized network analysis. Existing software, questionnaires regarding network analysis, software availability, and hardware requirements are discussed. An overview of the applications and some advantages of computer network analysis are presented.

Chapter 2 discusses activities and functions practiced in setting up network-analysis models. Data-collection methods and calibration are discussed. Network-data requirements, field measurements, water demand, and skeletonization of networks are explained.

Chapter 3 covers definition of the analysis problem, design criteria and other analysis considerations, the analysis process, and evaluation of water-distribution-system performance.

Chapter 4 discusses advanced topics, such as optimization of design and real-time computer-control techniques for operations, which have not yet become common engineering practice.

Chapter 5 describes in general terms computer-graphics routines used to enhance and expand the capabilities of standard network-analysis software.

* * *

This manual was prepared by the Computer Assisted Design of Water Systems Committee. Committee members involved in the preparation of the manual are as follows:

Lee Cesario (Chairman), Denver Water Department, Denver, Colo.
P.T. Brasher, City of Albuquerque, Albuquerque, N.M.
J.L. Buttle, Havens and Emerson, Cleveland, Ohio
C.D.D. Howard, Charles Howard and Associates, Victoria, B.C.
Joe Kroon, Stoner Associates, Carlisle, Pa.
E.F. Morey, Turbitrol Company, Atlanta, Ga.
M.P. Robinson Jr., Malcolm Pirnie, Newport News, Va.
Simsek Sarikelle, University of Akron, Akron, Ohio
Uri Shamir, Technion, Haifa, Israel
R.E. Topping, CH2M Hill, Montgomery, Ala.
J.P. Velon, Harza Engineering Company, Chicago, Ill.
Tom Walski, City of Austin, Austin, Texas
D.J. Wood, University of Kentucky, Lexington, Ky.

Several people participated in the development of this manual through coordinating activities, writing, and reviewing. Generally, each chapter was written by one person and reviewed by several others. Any comments, questions, or corrections should be directed to AWWA or to the CADWS Committee chairperson.

Chapter 1

Introduction to Network Analysis

Network analysis is the process of using a computer model of a water distribution system to analyze performance capabilities and to define the system requirements necessary to meet system design standards for pressure and flow. A computer model is composed of two parts: a network-data file that defines the system in terms of pipes and nodes and a computer program that solves a series of hydraulic equations for pressure and flow.

Network data include physical characteristics for pipes, such as length, roughness coefficient, diameter, and node data, which describe elevation and water demand or water supply. Nodes, or junctions, define the end points of pipes or pipe segments. Operational parameters, which indicate the actual value of system facilities, such as pump-station flow rate or reservoir elevation, are also included in the model.

Calculating pressure and flow in pressurized pipe systems is a time-consuming process that requires several iterations to arrive at a solution. A solution is a set of pressures and flows that simultaneously satisfies the energy and continuity equations for every link and node in the network. Calculations for a system containing a few loops can be done manually, but this becomes impractical when more than a few loops are involved. Computer programs to solve energy and continuity equations have been in use for many years, but only during the last several years, with the advent of the microcomputer, has computerized network analysis become widely used.

HISTORICAL DEVELOPMENT

Network analysis of water systems has evolved during the past 30 years from a time-consuming process done infrequently on small systems to a quick and easy process done regularly on systems of all sizes. Because network analysis requires solutions to mathematical formulas and calculations that are more easily done by computer than by hand calculations, the growth in the use of network analysis parallels the growth of the computer industry. Several events in the history of network analysis resulted from advancements in computer technology.

Fundamentals of Fluid Flow

In order to understand the historical development of computerized network analysis, one must first understand the fundamental laws of physics on which it is based. Two laws of physics that apply to water-system networks define flow continuity and head loss around a loop. Flow continuity means that flow entering a node must be equal to the flow leaving the node plus the water use at the node; in other words, the total flow must be accounted for. Head loss around a loop refers to the law stating that the sum of head loss in each pipe is equal to the total head loss in the loop, which is equal to zero. A loop is a series of pipes connected in such a way that the beginning of the first pipe is the end of the last pipe and no pipe is traced twice. These fundamentals of fluid flow are used in all network-solution methods. The Hardy Cross method of calculation, which uses a systematic numerical scheme to calculate flows that satisfy these physical laws, has been used since 1936.[1]

Manual network calculations were time-consuming and susceptible to error; this made them candidates for solutions using computer techniques. As computers made their way into the business world, engineers and other professionals began to recognize the possibility of using computers to solve network calculations.

1940s

The McElroy Analyzer was developed and used in the 1940s to analyze water distribution networks. The analyzer, using an analog computer rather than the now-common digital computer, used wires, resistors, and light bulbs to simulate a water network. Resistors simulated flow restrictions. The method was crude by current standards, but it did provide a physical model.

1950s

Network analysis was done on large mainframe computers in the 1950s. A few network-analysis programs were available, but their use was relatively limited. Some utilities developed their own network-analysis programs, as did the City of Dallas, Texas, in the late 1950s.[2] However, use of computers was limited by their expense, physical size, and the extensive programming efforts required to implement applications.

1960s

Several computer programs for network analysis were developed and implemented in the 1960s. Two programs that became widely used are the Shamir–Howard program,[3] developed in 1965, and the Epp–Fowler program, developed in 1969.[4] Both programs made use of the Newton–Raphson technique for calculations. However, there were two notable differences in the programs. First, the Shamir–Howard program was based on a node-continuity formula, while the Epp–Fowler program was loop oriented. Second, the Shamir–Howard program solved for a mix of unknowns, including pressure, consumption, and pipe and node characteristics, while the Epp–Fowler program solved for pressure and flow.

Other network-analysis programs were available on a time-sharing basis. Utilities and consultants accessed a mainframe computer through a teletype terminal. Files containing water-distribution-system data were created, the program executed, and the output printed immediately. Thus, computer calculations for large systems became possible. Using these programs, several cases could be analyzed relatively quickly and easily.

Some network-analysis programs developed in the 1960s used solutions (pressures, tank levels, and flows) for one point in time as input to the program for a second point in time, and then made calculations to result in a second solution. This set of multiple solutions represented the operation of a system over a series of time periods. It became known as multiple-solution simulation, continuous simulation, time simulation, or extended-period simulation.

With new technology came new problems and limitations. Two of the problems were the cost of processing time and the length of turnaround time. Users were concerned about the number of cases analyzed and the size and components of the system analyzed, because the costs of analyses were based on the number of runs and the amount of computer-processing time used.

Two solutions to the problems were available: running interactively or running in a batch mode. Running interactively was expensive because it required immediate use of the computer's resources. Interactive execution produced results quickly, but the expense could seldom be justified. Running in a batch mode, which allowed use of the computer during off-peak times (generally at night), was less expensive. However, when running in a batch mode, the user had to wait for results—sometimes for hours and sometimes until the next day, depending on the demands placed on the computer. Thus, errors in data files could not be detected immediately and analysis that depended on previous results was delayed.

Whatever method was used to run the programs, certain limitations were inherent in the programs themselves. For example, analysis was generally limited to single pressure zones and/or multiple pressure zones if pumps were used and pressure-regulating valves were not used. Analysis of more complex systems required execution of the program for each zone. Downstream pressures at pressure-regulating valves were used as input for analysis of another zone.

1970s

The 1970s brought enhancements to popular programs and development of new programs. This generation of programs had expanded capabilities based on improved mathematical and computer-programming techniques as well as advances in computer hardware. Three such programs are the University of Kentucky, Lexington, Ky., program, developed by Wood in 1972;[5] the Utah State University, Logan, Utah, program, developed by Jeppson in 1978;[6] and WATSIM, developed by Systems Control, Inc., Palo Alto, Calif., under a federal government contract in 1974.[7]

During this decade, universities began to offer network-analysis seminars that covered the fundamentals of fluid mechanics and provided hands-on use of programs. In most cases, participants were urged to bring network data to use in analyzing a real problem.

Significant improvements in the capabilities of network analysis began to appear in the 1970s. Programs could simulate all of the components of a water system, including pump stations, pressure-regulating valves, check valves, and reservoirs. Continuous-simulation capabilities were improved and more readily available.

Problems of cost, interactive versus batch processing, and turnaround time became less significant. More powerful computers resulted in faster processing, which made interactive processing cost-effective. Interactive processing eliminated problems with slow turnaround time, because results could be seen within seconds.

Support programs to aid analysis began to appear in the late 1970s. The Denver Water Department, Denver, Colo., developed programs for load allocation, network building, and output comparison in 1979.[8,9]

Also in the 1970s, computer-processing power expanded and the cost of hardware decreased. With less expensive computer costs, increased awareness of computers and their capabilities, and good software, computerized network analysis began to play a major role in the design of water systems of all sizes.

1980s

The microcomputer began to have a significant impact on network analysis in the early 1980s. Microcomputers of the mid-1980s were more powerful than large mainframe computers of the 1950s and 1960s and, therefore, were capable of solving complex mathematical problems. In addition, microcomputers were low enough in price that their use could be easily justified based on their potential to save money.

Popular network-analysis programs of the 1960s and 1970s were converted to run on microcomputers, and many new programs were introduced by software vendors and consultants. The Shamir–Howard, Epp–Fowler, Wood, and Jeppson programs were converted, and hundreds of programs became available through vendors, software publishers, and consultants.

Capabilities of network-analysis programs of the 1980s were greater than those of the 1970s. Simulation of all water-system components in static state and continuous period became routine. Attention began to shift to expanded capabilities, such as data-base management, network graphics, pumping-energy cost calculations, network building, data manipulation, and sizing-optimization routines. Data-input methods were made easier and output was easily manipulated.

A program introduced by Sarikelle of the University of Akron, Akron, Ohio, in 1979[10] combined computer mapping and data base with network analysis. The Akron program was developed on a mainframe computer, was converted to a minicomputer with supporting computer-graphics hardware and software, and in 1986 was converted to run on a microcomputer. Pressure contour plotting, improved fire-flow calculation techniques, and the ability to report on information in the data base were key features of this program. These features were not previously available in most other programs.

A water-distribution simulation and optimization program (WADISO) was developed in 1985 by Gessler and Walski.[11] This program, which is in the public domain, focused on pipe sizing and rehabilitation optimization for use by the US Army Corps of Engineers.

During this decade, several utilities began to look for methods of combining various engineering computer applications with network analysis, because much of the data for each is common. Mapping information management systems (MIMS), computer-assisted design systems (CAD), automated mapping/facility management systems (AM/FM), and similar systems using computer graphics and a facility data base were used to generate maps, drawings, and reports. These types of systems also provided capabilities to plot networks, pressures, flows, and other network-solution outputs. The Denver Water Department, City of Bellevue (Wash.) Water Department, and other utilities began to implement these types of systems.

APPLICATIONS OVERVIEW

Applications of network-analysis software generally fall into three categories: planning, design, and operations.

Planning

A primary planning application of network-analysis software is the development of long-range capital-improvement plans, which include scheduling, staging, sizing, and the preliminary routing and locating of future facilities. Other applications include the development of a main-rehabilitation plan and system-improvement plan. Rehabilitation plans call for cleaning and/or cement-mortar lining of mains. System-improvement plans call for installing new mains to keep up with growth or to upgrade the distribution system to utility standards. Network analysis and modeling can also be used to develop main-sizing guidelines or a standard grid to provide the desired pressures.

Design

Network-analysis design applications include the sizing of various types of facilities. Pipelines, pump stations, pressure-regulating valves, tanks, and reservoirs can be sized using pressure and flow calculations resulting from network analysis. In addition, system-performance capabilities can be analyzed to determine fire-flow capabilities and the improvements necessary to meet fire-demand requirements.

Operations

Applications to system operations include the development of operating strategies, operator training, and system troubleshooting. Operating strategies may be based on emergency conditions, energy-management restrictions, water availability, or other conditions of particular concern. For example, contingency plans for an outage of a key facility, such as a pump station, can be developed. Computer modeling and network analysis can also be used to develop operational strategies based on energy-management guidelines and restrictions for more efficient system operations. Strategies for shifting loads among treatment plants can be developed to maximize use of available water.

Modeling and network analysis are also good means of training personnel involved in the operation of distribution systems. Distribution-system operators can experiment with the model to determine how the system will perform under specified operating conditions.

System troubleshooting, based on modeling and network analysis, can be used to determine the cause of various problems, such as low pressure and unexplained events.

Network-analysis software is widely used in a variety of applications. The user must decide what applications, or uses, are most important in his or her situation. It should be noted that network analysis is a tool used to analyze situations; the user must exercise a great deal of judgment and interpretation to make use of the results.

ADVANTAGES OF COMPUTERIZED NETWORK ANALYSIS

Feasibility

The most important advantage of computer modeling is that it makes network analysis feasible. Without computerized techniques, analysis is slow and impractical, except for very simple systems. Before computerized network analysis, overdesign was the common reaction to uncertainty caused by the inability to accurately predict flows and pressures in a system. By providing reliable information, network analysis enables the user to more efficiently analyze, design, and plan water systems.

Cost

The cost of computer hardware and software is relatively small compared to the amount of money that can be saved by designing a water system efficiently and accurately. Water facilities are expensive, and the cost savings achieved through proper sizing and installation can range from thousands of dollars to millions of dollars. Efficient operating strategies developed through use of network analysis can also be used to minimize energy costs, thus increasing cost savings.

Socioeconomic/Environmental Impacts

Several socioeconomic and environmental considerations make network analysis a standard utility tool. Increasing construction costs, combined with increasing population and consumption, make systematic analysis necessary to efficiently design, plan, and operate a system. Water-consumption-management plans can be used to save water and to redefine consumption patterns that lead to more efficient use of facilities.

Familiarity

A final advantage of network analysis is that it enables the user to gain a feel for system capabilities and limitations. This is required to meet industry trends. Emphasis on water quality in distribution systems, fire-flow capabilities, and advanced system-operation techniques and tools make understanding a system increasingly important.

SOFTWARE

There is an abundance of network-analysis software. The AWWA Computer Assisted Design of Water Systems (CADWS) Committee has attempted to compile information about programs and their use. Several surveys and questionnaires by the CADWS Committee and other groups have been distributed during recent years (see Table 1-1).[12]

Table 1-1 Computer Questionnaires Distributed by AWWA CADWS Committee

Questionnaire Title	Date
Computer Assistance Group of Water Utilities of America (CAGWUA)—Questionnaire 1	July 1980
Engineering/Scientific Computer Usage Questionnaire (AWWA)	December 1980
Computer Assistance Group of Water Utilities of America (CAGWUA)—Questionnaire 2	February 1981
Network Analysis Survey (Denver Water Department, Denver, Colo.)	September 1981
Network Analysis Survey 1984 (AWWA–CADWS)	June 1984
Microcomputer Software Usage in Water Utilities (AWWA)	November 1984

Network Analysis Survey 1984

In an effort to learn recent national trends in the development and usage of network-analysis software, the Network Analysis Survey 1984 was conducted. The survey, which was designed, distributed, and compiled by the CADWS Committee, was mailed to 396 US water utilities. The questionnaire contained questions in three primary categories: general information, computer usage, and *C* factors. All responses are compiled in the Network Analysis Survey 1984 Report.[13] These results represent the most current information available at the time this manual was written.

There were 172 respondents to the survey. Of these, 109 used network-analysis programs. Responses to the survey indicated that many programs are in use and more programs were implemented in 1983 than any preceding year. Other significant results include the following:[12]

- 5 percent developed their programs in-house,
- 52 percent modified their models,
- 88 percent planned to improve their models,
- 40 percent plan to establish models in the next two years, and
- 92 percent felt their models were suitable in complexity.

Current uses of network analysis, beginning with the highest percentage, were: long-range planning (95%), fire-flow assessment (78%), operational strategy (52%), infrastructure rehabilitation (40%), emergency-response plan (25%), and energy management (17%).[14] Several papers written by committee members discussing survey responses have been presented and published at AWWA conferences.[14,15,16]

Trends in Network Analysis

Two key trends emerged from the Network Analysis Survey 1984. First, a parallel was noted between the increased use of network analysis and the introduction and use of microcomputers. This was determined by observing the brand names of the computers being used as listed in the results.

A second trend was the use of computer programs that perform more than the basic network-analysis functions of calculating pressure and flow. Comparison of responses of the 1981 and 1984 network analysis surveys indicate that many users have moved beyond basic network analysis to more sophisticated capabilities, such as energy calculations, network loading, and network building. (It should be noted that the 1981 and 1984 network analysis surveys were constructed and conducted by committee members without the aid of professional pollsters; nevertheless, the trends are quite clear and can be easily recognized.)

Software Capabilities

Many of the steps involved in network analysis can be performed by a computer with the proper software. Following is a brief discussion of network-analysis-program and support-software capabilities as identified in the Network Analysis Survey 1984.

Automatic hydrant use: The ability to automatically select the hydrants required to meet a specified fire-flow requirement in a given vicinity. The program determines the number of hydrants and their flow capabilities based on flow requirements and the number of hydrants to be used.

Demand forecasting: The process of using consumption values based on land use and consumption ratios to arrive at a consumption projection for a geographical area under given conditions, such as peak hour.

Design optimization: The process of determining pipe sizes and least operational cost based on selected sizes, cost per foot for selected pipe sizes, and power costs.

Energy-use calculations: The use of energy-cost factors to calculate energy requirements and pumping cost.

Flow: The calculation of flow in pipes in the network.

Graphics: The graphic display of network data and network solution output, such as pressures, flows, pressure contours, and other similar items.

Hydraulic grade line (HGL): The calculation of hydraulic-grade-line values at nodes. The HGL is equal to the pressure, in feet, plus the elevation.

Hydrograph generation: The creation of hydrographs for various conditions, such as demand and reservoir elevation. A hydrograph is a graph having a horizontal axis defined in units of time and a vertical axis defined in units of the item being measured, such as feet for reservoir elevation or gallons for consumption.

Network-data generation: The process of extracting selected network data from a data base and automatically formatting this data for input into the network-analysis program. This capability can be used to combine network-data files for different areas or to extract a network for a selected area from a larger area.

Network loading: The process of calculating and distributing loads or demands to the nodes in the network-data file. Sometimes also called load allocation or load distribution.

Network plotting: The ability to plot a network-data file using computer techniques; it is also called automatic plotting.

Optimization of operations: The process of determining the least-cost operation strategy based on operating costs, incorporating energy-use calculations.

Output comparison: The comparison of network-analysis output values for various cases. For example, a program with this capability can create a table that lists pressures at selected nodes for various demand conditions, such as peak hour or minimum hour, or with or without specified facilities. These computer-generated values can then be compared to actual values.

Ability to change parameters: The ability to change parameters and factors that are used in solving and defining the network. Parameters and factors include, among others, roughness coefficient, demand factors, units of flow, tolerance factor, and number of iterations for solving the network. Specific variables may include, for example, use of Hazen–Williams and/or the Manning formula for pipe-flow characteristics, use of a multiplicative factor to change all loads, and use of million gallons per day (mgd) and/or gallons per minute (gpm) for units of flow.

Pressure: The calculation of pressure at nodes.

Pressure contour plotting: The plotting of pressure contour lines (similar to ground contour lines).

Pressure/flow plotting: The ability to plot network-solution output (pressure and flow) automatically.

Simulation of standard water-system components: The ability to simulate standard components of a water system, such as pumps, pressure-regulating valves, check valves, closed valves, and reservoirs.

Time simulation: The calculation of pressure and flow in the network for varying demand conditions over a period of time; 24 time periods at 1-hour intervals is common.

One should keep in mind that there are numerous programs with a variety of capabilities currently available. An individual might consider developing simple data-manipulation and analysis programs, if such software is not accessible.

Sources of Software

Network-analysis programs are available from a variety of sources. Four sources are engineering consulting firms, software vendors or publishers, universities, and utilities and government agencies.

Consulting firms. Engineering consulting firms have provided network-analysis services for many years. Consulting firms frequently perform capital-improvement studies that require the use of computerized network-solution programs. After the network data base is created, the analysis performed, and the improvement plan provided, the consulting firm usually offers the utility a number of options for continued use of the network-solution program. The consultant might lease a copy of the program to the utility for in-house use. The consultant might also make the program available on a time-share basis, which allows the utility to perform its own analysis while sharing hardware. Another option is to have the consulting firm perform all analyses, in which case the utility has no control of or access to the analysis program. In all cases, the consultant maintains and upgrades the software.

Software vendors and publishers. Software vendors and publishers develop, maintain, and sell network-analysis software. Generally, these companies are primarily concerned with selling programs and providing software support, rather than setting up the data base and doing analysis.

Universities. Several universities have developed network-analysis programs that have been widely used and, in some cases, modified by utilities, consultants, and vendors. University programs are perhaps the most widely used programs, because they were, and still are, relatively inexpensive and are frequently updated. Seminars in the fundamentals of hydraulics and network analysis are held periodically by several universities.

Utilities and government agencies. A number of utilities and government agencies share their software with other utilities. In some instances, a nominal fee is charged to cover copying and mailing expenses. Some utilities also require an agreement that their software not be released to a company that might use it for profit. Programs in this category were generally created and placed in the public domain (such as WATSIM) or were developed by a utility. The latter programs were developed to meet the needs of a particular utility and, therefore, might require modification to meet the needs of another utility. While such programs appear to be free, some require a great deal of work to modify and implement; they may be poorly documented as well. However, other utility-developed programs may be both easy to use and well documented.

In choosing a software program, the potential user must evaluate the utility's needs and resources in order to select the most suitable program. User needs include program capabilities and the amount of customizing required; resources refer to hardware, number of people, and skills.

HARDWARE

Computer hardware requirements for network-analysis programs vary according to individual programs; however, certain computer hardware components are required by enough programs to warrant general discussion here.

Network-analysis programs are available for mainframes, mini-, and microcomputers. Mainframe computers are large computers with extensive capabilities. Mainframes perform many functions; network analysis is simply one of them. Minicomputers are similar to mainframes in that they have extensive capabilities;

however, they are generally used for specific types of applications, such as computer-assisted design, process control, or computerized mapping. Microcomputers are also used for many applications, but these applications are generally handled one at a time.

Standard Components

System unit. Standard hardware components for a microcomputer system include the system unit, a monitor or cathode-ray tube (CRT), a keyboard, at least one disk drive, and a printer. The system unit contains the central processing unit (CPU), which commonly has at least 640K bytes of random-access memory (RAM). Memory requirements of network-analysis software vary widely. The memory required by a program is some indication of the capabilities of the program. A program that requires 128K bytes is probably more limited than a program that requires 640K bytes. The 640K-byte program probably can handle a larger network, as determined by the number of pipes and nodes. The larger memory may also allow for use of a numeric coprocessor, which can greatly decrease the time required for mathematical calculations.

Monitor. Some programs make use of various colors, and a color monitor is required to make use of this feature. However, many programs written for monochrome, or one-color, monitors will also run on color monitors. Use of various colors can be extremely helpful in the analysis process but is not essential.

Keyboard. A keyboard is a standard component that provides a method of communication between the user, the system unit, the printer, and the monitor.

Disk drives. At least one disk drive is required to store the program and data. Generally, two disk drives are better and a hard disk is better yet. The use of two disk drives or a hard disk minimizes disk swapping and thus lengthens the life of drives and disks. Hard disks are also capable of handling large data files.

Printer. A printer is required to obtain a printout of network data and results. Two types of printers are available. Impact printers include the original dot matrix and daisy wheel. Nonimpact printers are known as ink jet and laser, although they might also use a dot-matrix technique. The quality of print varies among printers; dot matrix and letter quality are two common types of print. Dot matrix is generally of lesser quality than letter-quality print.

The graphics capability of a network-analysis program might require a printer with graphic capabilities. For example, if the program produces screen graphics, then a graphics printer that will perform a graphics screen dump is required.

Other hardware. Other hardware that can be used with some network-analysis programs includes a mouse (or puck), a digitizing tablet, a plotter, and a graphics board. A mouse or puck may be used to communicate spatial information to the system unit. Generally, a mouse is used with menus to select commands for execution. It can also be used with a digitizing tablet to establish coordinates and create network drawings. A plotter is used to create high-quality, accurate drawings. A graphics board for insertion in the system unit is required for programs with graphics capabilities; in some cases, the graphics board is included with the CPU when it is purchased.

Minicomputer systems have components similar to those of a microcomputer system. A digitizing table having a much larger surface area than a tablet is available. Dual-screen work stations and tape drives are also common.

Summary

Hardware requirements vary depending on the requirements of the network-analysis software. Many programs operate using standard equipment; however, some programs with extensive capabilities require additional hardware. In short, hardware and software must be evaluated jointly to ensure that the system can perform the desired capabilities.

References

1. CROSS, H. Analysis of Flow in Networks of Conduits or Conductors. Bull. 2 and 6. Univ. of Illinois, Engrg. Exper. Stn., Urbana, Ill. (1936).
2. CESARIO, A.L. *Network Analysis Survey 1981*. Denver Water Dept., Denver, Colo. (1981).
3. SHAMIR, URI & HOWARD, C.D.D. Water Distribution System Analysis. *Jour. Hydraulics Div.—ASCE*, 94:HY1:219 (Jan. 1968).
4. EPP, ROBERT & FOWLER, A.G. Efficient Code for Steady-State Flows in Networks. *Jour. Hydraulics Div.—ASCE*, 96:HY1:43 (Jan. 1970).
5. WOOD, D.J. & CHARLES, C. Hydraulic Network Analysis Using Linear Theory. *Jour. Hydraulics Div.—ASCE*, 98:HY7:1157 (July 1972).
6. DAVIS, A.L. & JEPPSON, R.W. Developing a Computer Program for Distribution System Analysis. *Jour. AWWA*, 71:5:236 (1979).
7. RAO, H.S. ET AL. *Extended Period Simulation of Water Distribution Networks*. Ofce. Water Res. and Tech., US Dept. Int., Contract 14-31-0001-0927 (Feb. 1974).
8. CESARIO, A.L. & LEE, T.K. A Computer Method for Loading Model Networks. *Jour. AWWA*, 72:4:208 (1980).
9. CESARIO, A.L. Computer Modelling Programs: Tools for Model Development. *Jour. AWWA*, 72:9:508 (1980).
10. FLASCO, R.D. Use of Computer Plotting in Hydraulic Analyses. Proc. AWWA Ann. Conf., Las Vegas, Nev. (June 1983).
11. GESSLER, JOHANNES & WALSKI, T.M. *Water Distribution System Optimization*. Dept. of Army, US Army Corps of Engineers, Washington, D.C. (Oct. 1985).
12. CESARIO, A.L. *Microcomputers for Water Utilities*, 79–83. AWWA, Denver, Colo. (1986).
13. Network Analysis Survey 1984. CADWS Committee, AWWA, Denver, Colo. (June 1984).
14. CESARIO, A.L.; MOREY, E.F.; & BRASHER, P.T. Computer Assisted Design of Water Systems. Proc. AWWA Ann. Conf., Dallas, Texas (June 1984).
15. VELON, J.P.; CESARIO, A.L.; & SHAMIR, URI. Network Analysis of Treated Water Systems. AWWA DSS Proc., Syracuse, N.Y. (Sept. 1984).
16. SARIKELLE, SIMSEK & CESARIO, A.L. Developments in Water Distribution System Analysis and Graphics Programs. AWWA DSS Proc., Seattle, Wash. (Sept. 1985).

Chapter 2

Preparing the Computer Model

Network analysis is the process of analyzing a water distribution system through use of a mathematical computer model of the water system. A base data file describing the network, operating parameters, and demands used in conjunction with a network solution computer program produces a model of the system.

Network data includes the physical description of the water network, including pipes, nodes, pump station capacity, reservoir capacity, and valve sizes. Operating parameters include actual system values, such as pump station flow rate, reservoir elevation, and pressures for a particular point in time, such as peak hour. Demands are water-consumption values for nodes for a specific time; a multiplicative factor can be used to adjust all demands uniformly to represent a different point in time. Key information output from the model include values for pressures, flows, hydraulic grade lines, and head loss in pipe segments.

To ensure that a model is accurate and reliable, it must be calibrated. Since much of the required data is unknown, either measurements must be taken or estimates must be used; thus network analysis might be considered as much of an art as a science. Calibration is the process of making minor adjustments to the input data, or base data file, until the results, or output, accurately simulate field conditions within an acceptable range of accuracy. The calibrated model becomes a benchmark that is used to develop additional models for various cases and for comparison to determine the effects of changes.

STEPS IN ANALYSIS

Analysis of a water distribution system is generally accomplished in the following four steps:

1. Collect data.
2. Prepare a base model.
3. Calibrate the model.

Prepare a model of the distribution system, using physical and operational data, including data collected from water-system records, personnel, and measurements taken in the field.

- Conduct a water-system inventory (Figure 2-2).
- Establish a base model using the physical data, including piping, pumps, control valves, and storage.
- Establish the demand conditions that are most limiting to the performance of system components. These conditions are usually maximum-day demand, maximum-day demand plus fire-flow demand, peak-hour demand, or replenishment.
- Initialize the model for the limiting demand conditions.

Calibrate the model in two phases: one phase for physical data and one phase for operational data.

- Collect physical data on key system features.
- Debug the computer model using the actual physical data.
- Establish a program for collecting operational data.
- Collect operational data in order to observe and measure system performance and characteristics under various demand conditions.
- Run the computer model at various conditions observed during the field-monitoring program. From these runs, calibrate the computer model and establish the level of confidence in the model's ability to predict system performance. Further investigations may be necessary to resolve major discrepancies.

Figure 2-1 Model preparation.

4. Use the calibrated model to analyze the distribution system and develop an improvement plan and operational procedures.

The first steps are performed in preparing the distribution-system model (see Figure 2-1). The fourth step, network analysis, is discussed in chapter 3.

The computer model is the foundation of network analysis. In order for the analysis to be effective, an accurate model of the distribution system must be established. For this reason, computer model preparation is a critical early step in the process of network analysis.

MODEL COMPONENTS

A distribution-system model is composed of the network-data file and the computer program, or software, that manipulates the network data and performs the necessary mathematical calculations. The network-data file is composed of physical data and operating data. Physical data describe the network configuration and water demands. The physical-data portion of the network-data file is called the base data file. In like manner, the portion of the system model based on the physical data in the base data file is called the base model. The base data file, when fully developed, becomes the modeled representation of the existing distribution system.

Operating data represent the actual values of system facilities, such as pump station discharge pressure, valve position, and reservoir elevation. Since data can never be complete and totally accurate, it is necessary to make adjustments to these data until the model (output) accurately simulates actual field conditions. Once this has been completed satisfactorily, the model is said to be calibrated. A calibrated

Inventory existing facilities to determine

- Pipe locations, diameter, length, type, age, and present conditions.
- Storage facilities, including capacity, location, availability, dimensions, water-surface elevations, and connections to the system.
- The design and operating features of service and booster pump stations, including capacity, system head, elevation, and related features.
- Control valves' or regulating valves' elevation, operation, pressure setting, and purpose.
- Service area and pressure zone boundaries.

Review system operating records and interview staff to determine

- Diurnal and seasonal treated-water production records.
- Service and booster pump flows and pressures.
- Clearwell and storage-level variations for various demand situations.
- System operation criteria for pump stations, storage, and source pumps.
- System operating pressures determined through past testing or monitoring and fire-flow or ISO (Insurance Services Office, Inc.) testing and rating.
- Power-consumption records, costs, and rate structures.
- Observed low-pressure areas indicating system deficiencies.
- Observed fire-flow capabilities.

Review water-consumption records to determine

- Total water sales (billing records).
- Service population.
- Number and types of service connections.
- Consumption by meter-book route and by major users.
- Peak factors.
- System diurnal demand curve.

Develop water demands using water-consumption information and population and land-use projections. Existing and projected water demands should be established for the total system, sub-areas within the system, and major users on the following basis:

- Average annual-day demand (ADD).
- Maximum-day demand (MDD).
- Peak-hour demand (PHD).
- Fire-flow demand (FFD).
- Minimum-hour demand (MHD) or maximum storage-replenishment rate (MSR).
- Other site-specific critical demand situations.

Establish background information for power-management and operational-improvement programs by

- Plotting pump characteristic curves versus system head conditions.
- Establishing the average maximum-day pumping rate from the developed diurnal demand curve.
- Determining whether system storage or auxiliary power can be used to reduce the peak pumping rate to or below the average of the maximum-day demand.
- Checking the piping arrangements to storage complexes for replenishment constraints.
- Examining the pump curves versus system-head conditions to determine optimum operation sequence based on system demands.

Figure 2-2 Water-system inventory.

model then becomes the benchmark case from which other cases can be developed and compared. Demands can be adjusted (perhaps increased) to represent future cases.

The network-data file as a whole, when combined with network-solution software that solves for pressure and flow, is the analyst's primary tool in performing system analysis.

MODEL FORMULATION

The one common requirement of all water-network analysis projects, regardless of their specific objectives or degree of detail, is the accurate mathematical representation of distribution-system hydraulic performance. The accuracy of a system model is defined by the correlation between computed and observed system performance over the full range of operating conditions. Pressures, flows, and reservoir-depletion rates in the system model must compare favorably to those in the existing distribution system in order for the model to satisfy established acceptance criteria.

Characterizing the System

In order to create an acceptable computer simulation of a distribution system, the analyst must first collect sufficient data to define or describe (1) the physical description of the system, (2) proper allocation of water demand throughout the modeled system, and (3) a good understanding of the operational characteristics and responses of the system. A detailed water-system inventory provides much of this information (see Figure 2-2). Accumulating an abundance of data, however, is not, in itself, sufficient. The data must, more importantly, be incorporated into the model in a manner that reflects both the actual operating conditions of the system and the actual, physical components of the system.

Base-Data-File Formulation

The base data file is the formulated data-input file that represents the *existing* water distribution system. The base data file includes both the current physical description of the distribution system, that is, pipe lengths, pipe diameters, pipe-roughness coefficients, pump curves, and so on, and the current water-usage, or demand, information. Initial preparation and periodic updating of the base data file to provide an accurate, current model of the system are essential.

Degree of detail. Until recently, steady-state system-analysis models were the primary means of evaluating system performance. However, in the 1980s, use of dynamic or extended-period-simulation (EPS) models increased dramatically. Extended-period-simulation models provide a more comprehensive, integrated analysis of a distribution system over a defined period of time (generally 24 hours) than steady-state models, and they also allow for evaluation of various system-operating rules and extreme demand conditions. However, EPS models also require a more detailed base data file than steady-state models. Consequently, the study objectives and intended future use of a system model determines the type of program to be used and hence the degree of detail required during initial development of the base data file.

Degree of accuracy. The degree of accuracy required of the base-data-file parameters and, consequently, the system model depends, to a large extent, on the distribution system's sensitivity to change. For example, if small demand or pump-curve variations create a radical change in system pressures or in source-pump performance, then overall accuracy, including accurate characterization of pipeline hydraulics, careful allocation of demands, and good representation of actual pump field characteristics, is more critical than it would be for a distribution system less sensitive to such variations.

Data selection. The abundance and variety of data available in various water distribution systems makes data selection very important. With the sophisticated water-system monitoring and data-base storage systems now commonly available, there is a tremendous temptation to use every piece of information that can be accumulated. This temptation should be resisted; only relevant, useful information should be included in the base data file. For example, it is not uncommon to find in sophisticated monitoring and storage systems usage data for specific meter routes separated by user category, with peak usage factors defining the usage patterns for each user classification. It may also be possible, using such systems, to identify daily and hourly peak usage by zone in a multipressure service system. Rather than incorporate all of this information into the base data file indiscriminately, the model preparer must select the data best suited to the system model and network-solution program to be used. To determine if allocation of demands should be by user category, peaking characteristics, pressure zone, or other demand-detailing approach, the model preparer must address several issues: (1) the degree of accuracy required to satisfy the analysis objectives; (2) permissible sensitivity of the model to changes in the base data file, and (3) if choosing to allocate demands by user category or individualized peaking characteristic, availability of a data-base management routine or adjustment format that can easily create the different demand files needed from the initial base data file.

The decision as to how much detail to include in the base data file must be made early in the development of the model. The decision depends, in part, on the network-solution program that will be used. Some programs offer unique features suited to fairly sophisticated demand-allocation techniques. Such programs may allow several peaking factors to be assigned to specific nodal demands, or specified demands may be held constant. Programs with unique features or sophisticated capabilities may require a more detailed network-data file than is required by a simpler program. Final program selection may be affected by the degree of detail in the base data file.

Calibration

The success of model preparation becomes apparent during calibration. (A model is calibrated to match field conditions.) Comparing model-generated values to actual system pressures for a known flow condition indicates the validity of initial data assignments. Minor adjustments are to be expected in order to achieve the degree of comparative accuracy required to satisfy the study objectives.

GATHERING DATA

The importance of understanding the operating characteristics of a water distribution system when preparing a model and conducting a computer-assisted analysis cannot be overstated. The physical and operational characteristics specific to a given water system require the analyst to become intimately familiar with every aspect of the system, including water treatment and supply points; distribution-system network; and residential, commercial, industrial, and other water-user groups.

As a first step in understanding the characteristics of a particular water distribution system, the analyst should discuss the system with the personnel responsible for operations. These discussions allow the analyst to obtain valuable information from those with first-hand experience in running the system. Operators usually have the best understanding of the strengths and weaknesses of their system, such as pressure-deficient areas, low daily flows, low fire flows, and operating policies for pumps and valves.

Physical Data

As part of these discussions, the analyst should conduct a field inspection to obtain a first-hand understanding of the system's function, layout, and operating peculiarities. This inspection will familiarize the analyst with the number, type, and condition of water-supply facilities, pumping facilities, and storage facilities; types and age of pipe and condition of the piping network; how system components perform over the range of demand; how components interface with one another; what limitations and deficiencies are evident; and any recent work performed to improve system hydraulic performance and efficiency. The location of high-use customers may also be observed during the inspection to aid in allocating demands for use in the model.

Operating Data

Features to be evaluated by the analyst when reviewing system operations include

- high-service and booster pumps, including sequence of operation, method of control, family of curves, pressure variations throughout the range of demands, and so on.
- storage usage, including when it is used, when it is available, when it is filled, controls on storage use or filling, and so on.
- closed valves, whether any valves are closed for service reasons or are seasonally opened or closed.
- system monitoring and controls, including such questions as how are components interfaced and controlled; what controls are used and how is data obtained for the controls; are pumps pressure regulated or flow regulated; how are measurements of pressure and flow made; are there any control valves, and how are they operated; how is the interface between pressure zones controlled, and so on.
- historic record of deficiencies, including location of observed problem areas, such as chronic low pressures.

Network Data

Having developed an overview of system operations and physical makeup through the initial discussions and field inspection, the analyst next must acquire detailed information on the major components of the system. Network data must be collected on the existing pipelines, nodes, pumps, storage reservoirs, pressure-zone boundaries and control elements, regularly operated valves, and emergency-response procedures. Figure 2-3 outlines the physical data typically required for model preparation; gathering this data is discussed in detail in the following sections.

PIPE NETWORK

All computer programs available for water-system analysis require detailed numerical information on the pipe network used to transport the water from the supply points to the demand points. Although each program requires a slightly different input format, the information required is similar, because the mathematical equations used are similar. Information must be collected for all pipes and then input

Pipe segments

- Assign identification number or name to pipe segments.
- Establish the length between nodes.
- Determine the diameter of pipe and roughness coefficient for each segment. Age and type of pipe are valuable pieces of information.

Nodes

- Assign identification number.
- Establish ground-surface elevation at each node or use elevation at hydrant outlet. Ground elevations may be established only at key pressure-monitoring nodes and estimated at all other nodes. Pipe centerline elevations might be used at critical locations.

Supply pumps and booster pumps

- Assign nodes or line segments to each facility.
- Determine multiple points on pump-characteristic curve for each pump.
- Define on/off control levels (system pressure or tank elevation).
- Establish ground-surface elevation for each pump.
- Identify and assign values (or adjust pump curve) for pump-station minor losses.

Control valves/Pressure-regulating valves

- Assign nodes or line segment to each facility.
- Determine downstream pressure setting for each pressure-reducing valve.
- Determine control pressures for flow-control valves.
- Establish ground-surface elevations for each valve.

Reservoirs (storage facilities)

- Assign node location.
- Define capacity, dimensions, flow rate, and operating range.
- Establish ground-surface and water-surface elevations for each reservoir.

Figure 2-3 Physical data required for base model.

as a base data file compatible with the network-solution program to be used. Generally, fewer pipes are needed to accurately simulate distribution-system hydraulics than actually exist in the system. (A detailed discussion of selecting the pipes to be used appears later in this section.)

Continuity

The first step in describing the pipe network is establishing a numbering system for the pipe segments and nodes and assigning an identifying number to each. After a numbering system is established and the physical data is entered, the continuity of the pipe–node model should be verified. Continuity is a check to verify that node and pipe numbers are not repeated and that the base data file is properly defined.

Pipe Diameter and Length

Information on the diameter, length, age, and type of each pipe to be modeled is required. Also needed is the location and elevation of the interconnection points, or nodes, of each pipe. The system's owner can readily provide much of this information with the use of scaled drawings, intersection maps, and construction documents. However, many recent improvements to the system may not appear on available maps and drawings. Discussions with system personnel are useful to ensure the data obtained is current.

Data on older systems, or portions of systems with old pipes, tend to be less complete than data on newer systems. When information is not available and system personnel are uncertain about the existing pipe characteristics for major network elements, a field inspection and, as a last resort, excavation may be needed to obtain the desired data.

Node Elevations

Elevation data required for nodes can sometimes be obtained from system maps or drawings. More often, however, elevation is approximated using topographic maps. For this purpose, topographic maps of the system area with ground elevation contours of 2 ft or less are desirable. Topographic maps with 10-ft contours, such as those provided by the US Geological Survey, may be used if better information is unavailable. However, the analyst must recognize that the calculated pressures generated by the model may vary from the real system by a few pounds per square inch solely because of inaccurate elevation information (1 ft corresponds to 0.43 psi). Calculated pressures may be further skewed by a few pounds per square inch because the elevations obtained from topographic maps are of the ground surface and not of the buried pipeline. Although corrections to the ground-surface elevation can be made to approximate the pipe depth, use of the ground-surface elevation as marked on the topographic map is a conservative approach and is often used.

Accuracy. Accurate elevation data is needed for all pressure-control points within a system because the network-solution program calculates flows and pressures system-wide based on these known points. Pressure-control points include all locations where system pressures are fixed or are open to the atmosphere, including reservoirs, clearwells for water-supply pumps, and pressure-regulated discharge points. Accurate elevations are also required where pressures are being monitored for model calibration; this ensures that a valid comparison of actual versus predicted pressures is made.

While elevation data for each node is needed to calculate system pressure, the degree of accuracy required can vary, depending on the specific location of the node and the objectives of the system analysis. Elevation data at locations other than pressure-control points affect only the calculated pressure at that specific location and have no effect on calculated flows. Consequently, the accuracy of elevation data for these internal node points is not critical to the overall system analysis.

Pipe Roughness

A numerical roughness value, based on the material, age, and condition of each pipe, is required as part of the pipe input data. Roughness coefficients can be calculated by means of various flow equations; the form of the roughness coefficient included in the base model depends on the flow equation used in the network-solution program.

Hazen–Williams *C* factor. The Hazen–Williams equation and associated *C* factor is the most widely accepted flow equation for water-distribution-system analyses. This empirical relationship was developed primarily for aiding in water-distribution-system studies. *C*-factor values are well documented for the various commercial pipe materials available. These *C* factors are valid for straight runs of new pipeline and do not account for "minor head losses" occurring at fittings, valves, and other appurtenances. Consequently, *C* factors provided by the manufacturer usually must be reduced to account for total system efficiency. Where a particular minor loss is significant (5 ft or more of head loss), a minor-loss factor should be introduced. In most network-solution programs, this is accomplished simply, by a *K*-factor assignment for velocity-head calculation.

As pipe ages, the *C* factor decreases. Approximations of *C*-factor value reductions that occur with age, due to changing roughness or effective flow-through diameters, appear in the professional and trade literature. Each distribution system's water quality and operations affect the *C* factor differently. Values in the literature should be used only as a guide.

Darcy–Weisbach and Manning equations. The Darcy–Weisbach equation and associated roughness value e, and the Manning equation with roughness value n, are used in some network-solution programs. While the Darcy–Weisbach relationship for closed-conduit flows is generally recognized as a more accurate mathematical formulation over a wide range of flow than the Hazen–Williams equation, field data on e values are not as readily available as are *C*-factor data for pipe-wall roughness characterization.

Detailed information on all three equations is available through other sources and, therefore, will not be presented here. The reader is referred to standard hydraulic textbook references for more detailed discussions of the various flow equations in use.

Determining pipe roughness. Determining appropriate roughness coefficients, regardless of the formulation used, is critical in developing a distribution-system model. Generally speaking, for a new system with noncorrosive water, it is probably sufficient to use near-new textbook roughness coefficients to characterize the pipe-wall roughness. However, this is not always the case. Standard values that appear in trade and professional literature may not always reflect actual system conditions. For this reason, flow tests to measure the flow rate and head loss through a clearly defined, isolated segment of pipe are often required to determine the existing roughness coefficients. As the age of a system increases, the variability in roughness coefficients also increases. Consequently, analysis of older systems may require a more extensive field-testing program to define actual roughness coefficients than a relatively new system.

If flow tests are required, a variety of pipe types, sizes, and ages should be tested. The pipes that are tested should represent each type of pipe in the network. Careful consideration should be given to potential influences on the test results, including partially closed valves and fittings, location of hydrants to be flowed, major users along the test section, test velocities, length of run, and method of differential pressure and flow measurement. Roughness-coefficient assignments for the entire system may then be extrapolated from the results of the testing program. Specific field-measurement tests that can be used to obtain *C*-factor information are included in this chapter's discussion of field measurements.

For projected future-year analyses, it may be prudent to reduce the roughness coefficient several points to reflect the effect of aging on system efficiency. This can be done using literature values as a guide or using historical data, if available. However, the relationship between the roughness coefficient and time is nonlinear, and the reduction in the roughness coefficient over the past pipe life should not be extended into the future at the same rate.

Unlined cast-iron pipes. Cement-lined iron, lined steel, concrete, PVC, RFP, and asbestos–cement pipe materials are now used in place of unlined cast-iron pipe. Older, unlined pipes that have not undergone major pipeline and relining programs tend to have low *C*-factor values ($C \leq 80$) caused by major tuberculation and flow restriction. Lacking specific date-of-installation data on old pipelines, it can be generally estimated that pre-1950 pipe was unlined (although lined pipe was first installed in the mid-1940s).

Other Pipe-Network Characteristics

Valves, fittings, and other appurtenances installed in pipelines can be identified and their characteristics included in the base data file. Pressure-reducing valves (PRVs), which maintain a specified discharge pressure at the valve, can be included in most computer models. The information needed for data input includes the valve location, ground-surface elevation, designation of the upstream side of the valve, and the downstream pressure to be maintained. If this information is not available, a field test to measure the downstream pressure setting of the valve (that is, the pressure at which the valve begins to open) may be necessary.

Check valves, which allow flow in only one direction, are also included in a base data file. The location of the valve in the system, flow direction, the diameter, and minor-loss coefficient must be designated.

Model Limitations

Software restrictions. Network-solution program restrictions pertaining to the pipe network must be identified. The number of pipes, nodes, pressure-reducing valves, check valves, and similar items that can be included in the base data file are normally limited by the capability of the network-solution program. These program limitations restrict the size of the pipe network that can be evaluated. Through programming changes, these limits can be adjusted; however, caution should be exercised to protect the original network-solution program when programming changes are incorporated. In addition it should be noted that processing time for a particular network-solution program may increase substantially as the number of network elements (pipes, nodes, pumps, storage facilities, and the like) increase. Because processing time costs money, budget constraints may also restrict the system size and detail to be modeled.

Hardware restrictions. The capacity of the computer on which the model is run may also affect the size of the system model. For example, a mainframe computer, at present, has a greater storage capacity than a microcomputer. Therefore, a larger system model can be accommodated and the system analysis can be processed more quickly. When embarking on system analysis, computer-related restrictions should be evaluated to determine if a reasonable distribution-system simulation can be conducted on the equipment available.

Fragment models. Systems may be analyzed by fragment models, that is, by separating the total system into distinct service areas or fragments. This may be done to circumvent limitations or to more intensively analyze particular areas, such as areas of serious hydraulic limitations, to take advantage of program features that allow more efficient analysis in smaller models. Using fragment models is especially useful for evaluating individual pressure zones in a multizone system. Fragment models can be built individually and then combined for a composite model of the complete system.

Skeletonization

Skeletonization is the process by which a water-system pipe network is stripped of pipelines not considered essential for the analysis. The purpose of skeletonizing a system is to develop a model that accurately simulates the hydraulics of the pertinent pipelines delivering water throughout the distribution system while reducing the complexity of the model and minimizing the time for analysis.

Study objectives determine, in part, the degree of accuracy required of the model; the degree of accuracy, in turn, affects the degree of pipe detail to be modeled. An oversimplified model with few pipes modeled may not identify problem areas related to low pressures and flows because the problem areas are simply not included. On the other hand, an extremely detailed model, with many or all pipes modeled, may indicate localized problem areas that are not present in the actual system. The latter false indicator results from the sensitivity of small-diameter pipes in localized areas to slight variations in demand allocation or incorrect physical data.

Guidelines for skeletonization. The system analyst is responsible for selecting the appropriate pipelines to be included in the model. While this selection is highly variable, it is defined, to some extent, by the study objectives. For example, models used for long-range planning and the development of major capital-improvement programs need not include small pipes. Models used for operations analysis and for evaluating the impacts of proposed developer expansion of a system, should include some of the smaller lines. Some general guidelines to skeletonization for planning are presented in the following list. Pipe size is provided as a general guide and not as an absolute.

- Major feeder mains (16 in. and larger) should be modeled because they represent the primary conveyance system of most medium-to-large pipe-distribution networks.
- Secondary feeder mains (10 in. and 12 in.), which transport water to residential areas, commercial areas, and other localized demand areas, are often included so that the pressures and flows available to these areas can be identified.
- Selected small pipes (6 in. and 8 in.) may be included to complete pipe loops and to allow a higher level of detail in areas where specific problems have been identified. All pipes of these sizes, however, are not generally needed to simulate system hydraulic performance. Normally, pipes that have very low flows relative to the overall system flows may be eliminated.
- Pipelines smaller than 6 in. that have relatively small flows compared with the other pipes modeled are not normally included in a model. Inclusion of these pipes will often dramatically increase the model's sensitivity to demand allocations in localized areas and create the false indicator previously discussed. While small piping may affect localized areas, these areas usually can be identified without the aid of a model.
- Other discussions of skeletonization are available in texts discussing specific programs of hydraulic modeling.

Pump Data

Source pumps and booster pumps should be modeled. A source pump supplies water to a distribution system from a treatment plant, storage reservoir, or groundwater well. A booster pump elevates the pressure on the discharge side of the pump but does not supply additional water to the system.

Pump Curves

In older systems, adequate pump data is often the hardest information to collect. Manufacturers normally supply pump curves for each pump unit at the time of installation, but these pump curves are based on tests conducted before the pump is shipped and often do not reflect actual field operating characteristics. In addition, these pump curves do not include head losses associated with the pump-station piping (inlet valves, bends, and similar items), which can reduce the pump's effective station-discharge pressure and discharge rate. Finally, the curves do not indicate changes in pump performance due to wear of impellers and maintenance practices, such as replacing or shaving impellers.

Nevertheless, the manufacturer's pump curves for each pump in the system should be obtained. A discussion with system-operating personnel may reveal factors that affect the original pump characteristics and pump curves. Additional information on flow and discharge pressures may be available, if appropriate meters are in place. This data can be useful in adjusting the manufacturer's pump curves with field results; it should be reviewed.

It is the analyst's responsibility to evaluate the pump data and to decide whether a representative system pump curve can be developed. As a matter of good practice, pumps should be field calibrated. In field pump tests, flow meters and discharge-pressure recorders are installed at a given pump in order to measure the actual suction and discharge pressures at various flow rates. This allows construction of a pump curve that represents actual pump performance. After acceptable system pump curves have been developed, the data may be entered into the base data file as curve coefficients.

NOTE: The discharge-pressure recorder is usually located in an easily accessible location on the discharge piping. The analyst must be aware of this location in order to determine whether pump-station head losses should be included separately in the model or if the test pressure already includes pump-station head losses.

Other Pump Models

If it suits the study objectives and pump characteristics, pumps may be modeled as constant-head reservoirs (variable-speed pumps with set discharge pressure) or constant inflows. This eliminates the need for pump curves. Consequently, the form by which the pump characteristics are input to the model depends on the network-solution program to be used and the study objectives.

Reservoir Data

Reservoirs include both elevated and ground-level storage facilities that supply water via gravity or via a pump station. Water levels in reservoirs fluctuate with system demand. Reservoir capacities, tank dimensions, operating ranges, control valves, and service areas to be supplied are all important elements to be defined for base-data-file input. For steady-state analysis, much of this data is not included in the model, but is required for use in interpreting and evaluating analytical results. When extended-period simulations are conducted, reservoir data is included in the model so that water-level fluctuations can be calculated and altitude valves can be modeled correctly.

The capacity of a reservoir can be represented either as the total volume of water stored or as the available volume stored. In many cases, the available volume is less than the total holding capacity. System pressure restrictions and fire-protection and emergency-storage reserve volume requirements may decrease the volume of water actually available during peak usage periods. The available volume depends on the system condition being analyzed; it is discussed in detail in chapter 3.

Reservoir geometry, overflow elevation, and freeboard constraints, if applicable, should be obtained for each storage facility. The water-level elevation that indicates an empty tank or the minimum limit for available storage should also be identified. Control information that activates pumps as tank water levels change and defines the location and pressure settings of altitude valves is also needed.

Based on tank dimensions, a flow rate into or out of the reservoir can be related to an increase or decrease in the water level. This relationship is important in evaluating whether a storage reservoir has sufficient capacity to meet a specific need (such as fire flow, peak-demand flow, and similar demands) for a defined time period. For extended-period-simulation programs, this information must be included in the input file so the system model can continuously adjust the tank water levels based on the calculated flows into or out of the tank for each time increment specified. Most network-solution programs require specification of the operating range as well as geometric information or volume-versus-water-level information for each tank. For steady-state analysis, the analyst can calculate the change in storage volume using a volume-per-foot value for each reservoir.

FIELD MEASUREMENTS

Field measurements are taken as part of data collection. Taking field measurements involves testing and recording the performance of various system components. Measurements include recording values for pressure, flow, water consumption, and pipe-roughness coefficients. These measurements are used as a supplement to and first check of information provided by water-system records and personnel during the initial stages of data collection.

Equipment Accuracy

The results of system analyses are of little use if the base data file does not accurately represent the water system being analyzed. Using field measurements to verify information provided by water-system personnel and to collect additional pressure and flow information are key steps in developing a representative model. Inspecting the system's metering and recording instruments to determine their accuracies and taking field measurements with calibrated equipment to obtain information not readily available are also essential.

Inspecting the system's flow and pressure meters and reviewing available meter-calibration data allows the system analyst to evaluate the reliability of information gathered from these meters. If periodic maintenance and calibration programs are in effect, data obtained from existing meters probably will be acceptable. Many system owners or utilities schedule annual calibrations of all master meters. On the other hand, if meters are visibly damaged or unkempt and have not been calibrated for years, their readings may not be reliable. Most existing meters can and should be calibrated if a question of accuracy arises. When over- or under-registration is verified, adjustments should be made to data for the period in which registration errors occurred.

Flow and Pressure Tests

A variety of field tests can be conducted throughout a water distribution system to aid in developing good data on system performance.

Loss-of-head tests. Loss-of-head tests are used to determine friction-loss characteristics through a variety of pipe sizes, types, and ages. Roughness coefficients thus developed are included in the flow equations to calculate pressure losses through the system.

In a loss-of-head test, a segment of pipe, ideally of one diameter and straight, is isolated from the system. Isolation of a pipe segment in this manner allows a roughness coefficient to be calculated based on a single pipe size and type, rather than on a composite system of two or more different pipe segments. All main valves, with the exception of one end inflow point, are closed. A hydrant with a flow meter attached to the small nozzle is opened on the downstream end of the pipe segment to provide a measured flow through the pipe segment from the inflow point. For a more accurate flow measurement, a Pitot tube can be installed instead of the hydrant gauge.

Another method of testing for loss of head, the two-pressure-gauge test method, requires measurement of the residual pressures at two locations along the pipe segment. The pressure gauges used in this test are often connected to fire hydrants. The elevation of each hydrant should be surveyed to establish a relative pressure head. The accuracy of the elevation can be checked, when there is no flow in the pipe, by calculating the difference of the two gauges' pressure readings; the difference should correspond to the difference in elevation of the pressure gauges. By measuring the difference in pressure between the two gauges, distance between the two gauges, diameter of the pipe, and flow rate through the pipe, a roughness coefficient can be determined.

Problems in conducting loss-of-head tests can take several forms. Isolation of a pipe segment can be difficult if there are few valves in the system, if valves are old and cannot be closed, or if many bends, fittings, or other obstructions that create cumulative minor losses are present in the pipe segment tested. For larger-diameter pipe, a high flow rate is generally needed to create a measurable pressure difference between the measurement locations. In some instances, several hydrants, each with a flow gauge attached, may be needed to create sufficient flow through the pipeline. However, caution should be taken so that the flow velocity does not exceed 10 ft/s, because the validity of the Hazen–Williams equation is questionable at extremely high velocities.

Larger pipes may pose other problems in conducting loss-of-head tests. In a large pipeline, a short segment of pipe and velocities less than 2 ft/s may not provide the amount of head loss needed for an accurate measurement. The length between the pressure-measurement gauges should be sufficient to allow for a measurable amount of head loss through the pipe. It is necessary to generate enough head loss so that pressure differences can be read within the accuracy of the gauge. It is very difficult to find field gauges that can measure the few feet of head loss common in large pipelines.

The hydrant-flow method of determining head loss is generally limited to pipe sizes 12 in. or smaller. In some cases, it is important to estimate C factors in larger pipes that are of different material or age than the smaller pipe in the system. In these instances, loss-of-head tests can be run using a Pitot rod, a manometer, and a length of tubing connected to the Pitot rod and a downstream fire hydrant. The connecting tubing allows a direct measurement of head differences between the two connecting points. The advantage of this method is that it allows a more precise method of measuring small head losses by eliminating the necessity of subtracting less accurate pressure readings on two gauges.

Generally, maximizing the head loss that is caused *only* by the roughness of the pipe segment is an important consideration in a head loss test. A field observer with experience conducting such tests and who is capable of making field decisions to adjust the planned testing procedure can often make the difference between a successful and a marginally successful test.

Fire-flow tests. Fire-flow tests indicate the ability of a system to provide water for fires in a specific area. Fire-flow tests also can be used to aid in calibrating a model. The criteria used to judge the adequacy of a system for fire flow is based on Insurance Services Office (ISO) information. The ISO designates minimum flow recommendations for areas within a system based on the activities and structures in the areas. For example, a residential area generally has lower fire-flow needs than commercial, industrial, or public-use facilities.

In a fire-flow test, the discharge flow rate from a designated hydrant is measured while maintaining a residual line pressure of 20 psi. Interpolation of data may be required to determine the actual flow available at a residual pressure of 20 psi, if a different residual pressure measurement was taken during the test. Generally, a good-quality gauge, such as an ANSI Standard Type A gauge, or better, with a large face, is satisfactory. It should be remembered that a fire-flow test is only as accurate as the gauge used.

Fire-flow tests indicate the distribution system's ability to deliver water at a given point, at a given time, under current demand conditions. They do not indicate what is occurring in adjacent areas. Some adjacent areas may experience unacceptably low pressures during fire-flow tests. The tests also do not indicate the system's ability to provide water for fires when demands are more limiting than those used during the test, for example maximum-day demand or peak-hour demand.

To use the fire-flow data for calibration, system-wide pressure-control points should be recorded during the fire-flow test. Pressure-control points include water levels in reservoirs and tanks, operational status of pumps, pressure settings at pressure-reducing valves, and similar information. If sufficient data is obtained, a comparison between actual and predicted pressure, using fire-flow tests to calibrate or verify a steady-state model, can be very effective.

Other flow and pressure tests. Occasionally, two types of flow tests are conducted to determine 24-h flow patterns for specific pipelines or service areas. These field measurements aid in the development of water-demand data.

Transmission-line flow tests require a continuous-recording Pitot tube or flowmeter installed in a major feeder (trunk) main of a system. A 24-h record of the flow is used to create a flow-versus-time curve, which can be used to develop system peaking factors.

A district-pressure-zone inflow/outflow test, or water audit, can also be used to develop area demand curves over a 24-h period. This test is done by isolating a specific service area through valve operations and supplying the area by a limited number of mains, preferably one main. Discharge from the area is also through a limited number of mains. The difference between the inflow and outflow from the area is equated to the demand in the area, including metered and unmetered usage. Based on the results of a 24-h record of inflow/outflow characteristics of an area, area demand peaking factors can be established.

WATER-DEMAND DATA

As a final step in data collection, data regarding water usage is collected and input into the base model. In order to prepare an accurate representation of system performance, water demands must be allocated geographically and diurnally.

Geographic Allocation

Geographic allocations can be made using billing records, service populations, meter-book routes, and records of major users. Geographic allocations may be established by developing system maps of customer types and concentrations or by installing permanent or temporary meters on major transmission pipelines. Water demands vary with the type of development or customer; that is, demand is not a constant rate per square mile or acre. Customer type may include industrial, commercial, governmental, institutional, residential, recreational, and others. Just as water demand is not a constant rate per square mile, residential demands are not a constant rate per capita. They vary with the type of house or multiple-family dwelling. For example, houses with large lawns typically use more water on a per-capita basis than multifamily dwellings use. Therefore, the allocation per capita is greater for an individual living in a house than for an individual living in a multifamily dwelling.

Some utilities have implemented computer record systems that include the geographic location of each customer. If the water-distribution-system computer model is digitized, the system records can readily be interfaced with the computer model to establish water demands in a given geographic area. Without a digitized distribution-system model, the geographic demands could be obtained and manually input to the model.

Diurnal-Demand Allocation

Diurnal-demand allocation indicates demand variations throughout the day. This information must be obtained to establish peak-demand factors. Diurnal-demand characteristics can be expected to vary by geographic area. For this reason, the peaking factors over the range of demands must be input to the computer model for each node in the model. However, it is not practical to develop a peak factor for each node in the system. Therefore, peak factors usually are established for development types or service areas and are input for all nodes representing corresponding types or areas.

Several approaches may be used to develop diurnal demands and peak factors. The approach used with any particular system depends on its size, complexity, and unique characteristics. (One approach, for use in steady-state analysis, is described in chapter 3.) In all cases, diurnal-demand curves should be developed for major geographic areas within the system; where possible, these geographic areas should be established to coincide with development types. Diurnal-demand curves should also be developed for major customers, such as major industries and bulk-water purchasers.

Developing a diurnal-demand curve. To develop a diurnal-demand curve for the overall system, recordings must be made of water-source flows (that is, metering of water treatment plant and/or groundwater-supply pumping to the system) and of storage inflow/outflow. A net hourly demand would then be determined by first calculating net storage inflow/outflow for that hour and then adding net storage outflow to, or subtracting net storage inflow from, the water-source flows (see Table 2-1). A similar approach should be used when developing diurnal-demand curves for particular geographic areas, using metered flow into the specific area in place of water-source flows.

Another approach used to develop diurnal-demand curves is to identify typical diurnal patterns for types of land uses, such as low-density residential, high-density residential, and commercial/industrial/institutional. This can be done by isolating a homogeneous land-use area and monitoring flow over a period of time, typically 24 h.

Table 2-1 Typical Diurnal-Demand Calculation Format

Time	Source Flows	Storage Inflow/Outflow* Tank A	Tank B	Tank C	Net Demand
12–1 a.m.	450	(40)	(20)	(20)	370
1–2	450	(80)	(40)	(40)	290
2–3	450	(120)	(60)	(60)	210
3–4	450	(140)	(70)	(70)	170
4–5	450	(150)	(75)	(75)	150
5–6	450	(140)	(70)	(70)	170
6–7	450	(80)	(40)	(40)	290
7–8	450	(20)	10	10	450
8–9	450	30	15	15	510
9–10	450	30	15	15	510
10–11	450	40	20	20	530
11–12 noon	450	30	15	15	510

*Numbers in parentheses indicate inflow.

In this way, typical diurnal-demand curves for each type of land use can be established. This information can be applied to any node in the model, based on the land-use makeup of the area represented by that node.

It should be remembered that diurnal patterns, especially for residential usage, vary significantly with season. If nonsummer demands are important, then diurnal-demand curve measurements should be made for nonsummer periods.

When there is limited capability to isolate typical neighborhood areas, water-demand relationships established for nearby utilities in the same climatic region may be used. The Linaweaver housing density and value relationship may be used to adapt the water-demand relationships from the nearby utility to the utility under study.

A typical composite-system diurnal-demand curve for a major southeastern US municipality on a high-demand summer day is shown in Figure 2-4. Note that the diurnal-demand curve is generally plotted as percent of average flow versus time, not actual flow rate versus time.

Service-Area Data

In addition to the peaking factors established and assigned using diurnal-demand curves, current water-system demands must also be defined and allocated to specific demand-withdrawal, or nodal, points. The nodal demands included in the base data file simulate the actual system water usage by describing, quantifying, and allocating the usage of water throughout the modeled distribution system. These nodal demand assignments can be confirmed using historical records of average daily water usage (steady-state calibration) and 24-h water usage (extended-period-simulation calibration). Per-capita usage, both for the system as a whole and by user categories in service areas, provides the basis for defining current water usage in a system. Historical population growth rates, historical water-consumption rates, and the classification of customers (that is, residential, multifamily, commercial, or industrial) should be evaluated to develop representative nodal demand distribution.

Historical population data. Historical population data can be obtained from a variety of sources. US census data, collected every 10 years, gives a good historical summary of the growth of a community. Additionally, many states, counties, and cities have planning departments that supplement the census data, thus providing more

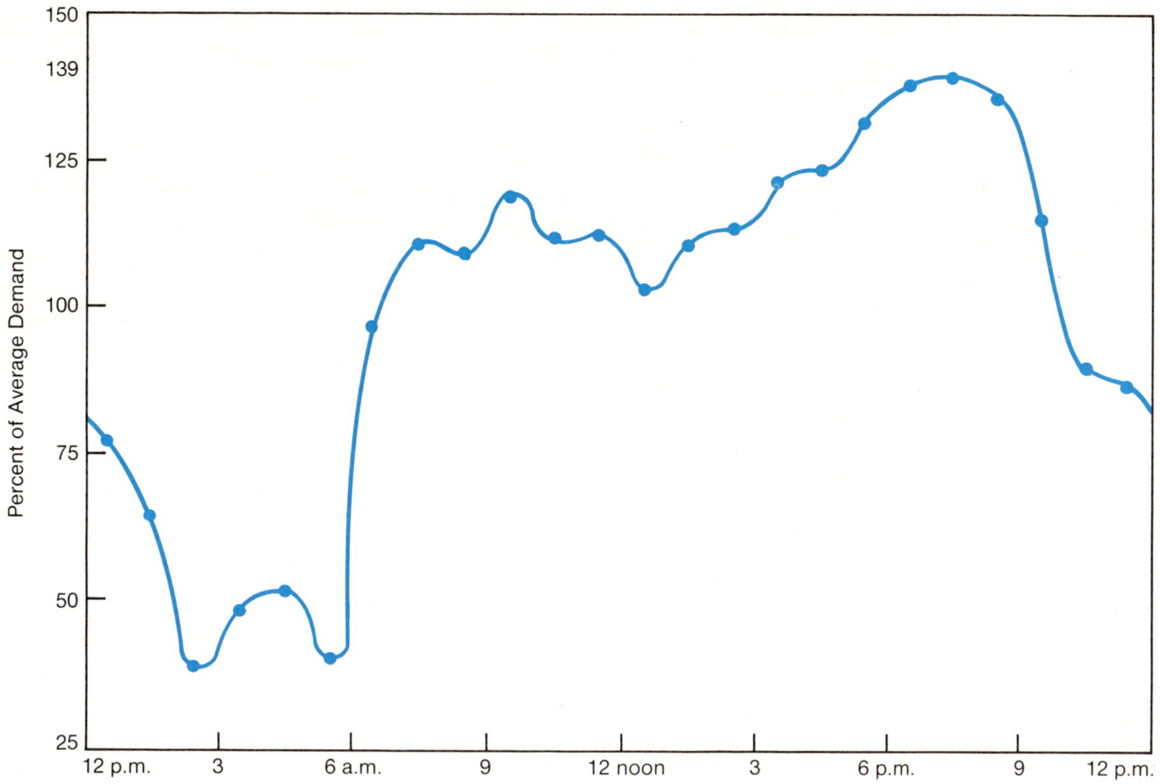

Figure 2-4 Typical system diurnal-demand curve.

detailed population data. The data available from these sources should be reviewed by the analyst to develop a clear understanding of the past growth of the service area and the effect of this growth on water usage.

The percentage of the total population that is served by the water system must be determined. In many locales, particularly in older areas, groundwater or other sources may be a source of supply to individual houses or developments. Where alternative water supplies exist within a service area, the total population data should be reduced to include only those users served by the water system.

Water-consumption rates. Per capita consumption is based on both the corrected population count and the average annual consumption patterns of the customers served. Meter-book records, location maps, and 24-h district zone measurements provide the basis for developing consumption rates. Billing records can be an excellent source of information to determine the actual billable consumption of a particular customer group. This data is usually the most organized and complete information available with regard to water usage. While some unmetered systems bill all customers the same monthly or quarterly user fees, this policy is not preferred and is gradually declining. Usually, each customer or group of customers is billed based on readings from a meter at a specific withdrawal point. A meter-route book is kept to record the monthly, bimonthly, or quarterly consumption of each customer.

Customer-information records vary widely from system to system. The analyst must judge the level of detail to be used. Monthly, quarterly, and per capita consumption records are generated for some systems as a whole, for specific user categories, or for specific geographic service areas. Consequently, the analyst must understand whether the per capita data generated is for the entire system, including all user groups, unaccounted-for water, and so on, or whether it is specific to particular user categories. With more user- and area-specific data comes the opportunity for developing a more accurate model.

Classification of customers. High-volume users, such as industrial, institutional, and heavy commercial users, are often readily identifiable and can be segregated from the lower, more evenly distributed demands of residential and light-commercial users. If the high-volume users are not isolated and defined by their usage characteristics, the resulting per capita consumption rate would be a more conservative estimate, reflecting a higher rate. This conservative estimate may lead to modest overdesign of a system-improvement program.

Other user categories, depending on the information available, may also be isolated and assigned individually designated demand rates.

Customer Categories

Customer categories represent groups of users with similar consumption characteristics. The number of categories for a given system depends primarily on the type of billing or metering information recorded. For example, the size of meters, if designated, could be used to separate user groups. Categories often designated in meter records include multifamily residential, single-family residential, commercial, industrial, and institutional. The residential and commercial categories are sometimes divided further into small, medium, and large services.

Multifamily residential units, including apartments, condominiums, and the like, generally have a slightly lower per capita consumption rate than single-family residential units. This is because multifamily housing characteristically has fewer people per household (that is, fewer children), the residents have more active lifestyles (and hence spend less time at home), and less water consumption for lawns and washing cars is needed. However, both categories generally have similar water use patterns over any given day, with the peak consumption occurring between 5 p.m. and 10 p.m. during the summer.

Commercial use consists of stores, restaurants, gas stations, offices, and so on. Commercial establishments have fairly consistent daily usage rates and predictable fluctuations over any given day. Major office complexes have slightly different consumption patterns than retail and service outlets, such as shops and restaurants, because of differences in the hours of operation. Therefore, office complexes could be separated into an individual category, if sufficient data is available.

Industrial clients are often high-volume users with specific and identifiable withdrawal points. These customers are often metered individually and billed on a monthly basis, so that accurate information is available. Characteristically, industries have very well-defined daily consumption patterns spanning the hours of operation. Industrial clients usually use water at a more steady, uniform rate than smaller residential consumers.

Government and institutional facilities, on the local, state, and federal levels, may also be designated as a user group. Government office buildings may be established as a separate user group or may be included under the commercial user group, as data dictates. Large military bases, prisons, hospitals, universities, golf courses, and similar institutions should be kept separate from other groups, because such facilities have their own consumption patterns related to the nature of the facility.

Unaccounted-for Water

Unaccounted-for water usage is always present in a water system. The unaccounted-for usage is estimated by comparing the average annual water production with the average annual metered consumption of a system. The difference between the two values is unaccounted-for water.

Unaccounted-for water usage can result from many factors. Some of the most prevalent factors include unidentified leaks in a pipe network, main breaks, periodic fire-hydrant flushing, tank drainage for maintenance purposes, unauthorized use, unmetered services, inaccurate and nonfunctioning meters, and water and wastewater treatment plant use. The uses of water at a plant for backwashing filters, mixing chemicals, rinsing equipment and tanks, and sanitary purposes are sometimes not metered and can represent up to 5 percent of the production rate for a system. Losses at the source or treatment facility customarily do not affect the model, as long as pump-suction characteristics are properly defined.

The unaccounted-for water usage must be added to system demands in the system model so that total water supplied will equal total water demand. The unaccounted-for usage is generally distributed equally to all nodes because specific or isolated causes are difficult to pinpoint, unless district zone measurements are made throughout the distribution system. System-wide district zone measurements permit a more accurate allocation of unaccounted-for usage. To increase accuracy, some systems have used leakage tests in subareas of the distribution system to prorate the unaccounted-for water usage. When, through subarea leakage tests, it was determined that various areas had various rates, the total leakage was allocated accordingly.

It is important to note that much of system analysis is conducted using peak-hour conditions. This reduces the impact of inaccurately distributing leakage to system nodes. For example, if total unaccounted-for water usage is 15 percent at average-daily demand, then at maximum-day demand it will generally constitute less than 10 percent, and at peak-hour demand, less than 5 percent. Such inaccuracy is generally less than the achievable accuracy of the model demand allocation.

The percentage of unaccounted-for water can vary widely from system to system. Values ranging from 4–30 percent of the total accounted-for consumption are found, although 10–15 percent may be more prevalent. The percentage can also vary from year to year in the same system. The higher values generally are associated with older systems, in which leakage, no meters, or faulty meters are more commonplace than in newer systems. Systems operating at high pressures usually will experience a high loss percentage.

Demand-Allocation Process

Demand allocation is the process of assigning water-consumption data to appropriate nodes in the system model. Consumption data from meter-route books or other sources are allocated to the nodes that best represent actual system withdrawal characteristics. Allocating demands to nodes is more an art than a science and requires, more than anything else, good working knowledge of system usage. Demand-allocation subroutines are available with some network-solution programs. A tabular approach, using a personal computer and spreadsheet software, can be an effective tool for expediting demand assignment.

Meter-route books. Meter-route data is of great value for allocating water consumption over a computer-simulated pipe network. Information available from meter-route books generally includes quarterly consumption for each customer and each meter route, location of each meter route, and identification of the user category for each customer within the meter route. This information can be readily developed into annual average usage data, which is then allocated by district. NOTE: The selection of a representative 12-month period, or normal use pattern, is extremely important for characterizing average annual usage. In some cases, only a seasonal, high-demand period may be determined, depending on study objectives.

Methods for allocating demands. There are several methods for allocating demands over a modeled pipe network. The approach described here uses a spreadsheet, meter-route map, and pipe-network map. The meter-book consumption for each route is listed across the top row of a table. All of the nodes to which a demand can be allocated are listed in the first column of the table. By laying a meter-route map over a map of the pipe network being modeled, the model nodes within a given meter route can be identified. Using consumption data already compiled, percentage allocations of the consumption can be made for each node within the route boundaries. The percentage value is recorded in the table. Special point loads, for industrial or wholesale use, are generally superimposed on the spreadsheet allocation of residential and commercial use. The demand to be assigned to each node can then be calculated. The calculation requires, for a given node, the summation of the portion of consumption for each meter route allocated to the demand node.

After the initial allocation is made and the spreadsheet assignments are complete, higher-than-normal demand periods or future growths in demand can be calculated by the application of multipliers to the data-file matrix. Initializing the model for the limiting demand conditions and use of demand files in system analysis are discussed in chapter 3.

MODEL CALIBRATION

After the physical data and the limiting operational and demand conditions have been established, and before performing the system analysis, the model must be calibrated. To do this, field data is collected and used to compare the model predictions of distribution-system performance to actual observed and measured system conditions. Without calibration, there is no way of knowing how accurately the model represents and predicts the performance of the actual system. Calibration thus establishes the level of confidence in the model's predictive capabilities. In general, a model would be considered acceptably calibrated if it predicts performance within 5–10 percent of observed performance. Many system models predict well within 5 percent of observed performance.

Field Data Collection

Field data collection should be accomplished in two phases (see Figure 2-5). In the first phase, system physical data is collected to confirm or establish data in the base data file. In the second phase, system operational data is collected for calibration. Physical data should be collected when demands are low to minimize interference in system operations. Operational data, on the other hand, should be collected during periods of greatest demand. The model calculates head losses at specific flow or demand conditions. At periods of greatest demand, the system experiences the greatest system head losses. Thus, more effective field measurements can be made of operational conditions because there is more head loss to measure at higher flows. The operational data should, therefore, be collected during or approaching maximum-day demand conditions. It may also be necessary to create high-demand conditions by conducting fire-flow tests.

Physical data collection

- Testing of supply pumps and booster pumps to establish curve characteristics.
- Loss-of-head testing to determine pipe-roughness coefficients.
- Pressure testing on fire hydrants, including areas of high localized demands and known low-pressure areas.

Operation data collection

- Continuous 24-h flow recordings at selected key locations, such as water production and major trunk mains.
- Hourly flow monitoring on additional key trunk mains and for large-demand customers.
- Continuous pressure monitoring at all continuous-flow recordings and key fire hydrants. Instantaneous pressure monitoring at key locations, such as fire hydrants.
- Continuous storage-level (hydraulic-elevation) and influence (inflow/outflow) monitoring.
- Continuous monitoring of high-service and booster pump operation, discharge flow, and pressure.

Figure 2-5 Field data for model verification.

Physical data. Field data is collected to verify information in the base data file regarding pump performance, roughness coefficients, and pressure in the system. Hydrant tests are used to make spot checks for pressure in various areas of the system. However, information obtained from hydrant tests should be recognized as a limited indication of system capabilities for several reasons. A fire-hydrant test measures pressure and flow-delivery capability at a specific hydrant in the system at a particular point in time. It does not recognize or indicate whether this is a limiting condition. It also does not indicate storage status (that is, elevation, fill demand, or output), demands of adjacent users, status of or relationship to the diurnal demand, and other related factors. Further, surrounding areas may have lower pressures than the pressures at the hydrant being tested because the ground elevation is lower, because of localized demands, because of piping limitations, or because of other factors.

To verify the accuracy of the base data file, demand conditions are varied. The base model is run at the average-day demand and maximum-hour demand to debug the model and to obtain some initial predictions of distribution system performance. At this point, a functioning mathematical computer model of the distribution system is established.

Operational data. The field program for collection of operational data should be developed based on the information gained during the system inventory, physical data collection, and base-model preparation. It is important to note that the computer base model should be prepared first and used as a basis for establishing the operational field-data-collection program. This provides for a limited field investigation effort that concentrates on key features and areas of the system.

Operational data should be collected over a short period, typically one to two weeks, during relatively high demands. The purpose of collecting operational data is to provide measurements and observations of actual system performance under specific demand conditions. The operational data should be collected to provide information on each key system component at each point in time, similar to a series of photographs. The distribution system should be divided into major geographical areas for monitoring demand disaggregation and diurnal variations.

Analysis of Calibration Runs

The collected operational data is used to calibrate the computer model. The base model is initialized with the observed monitored conditions, and the computer model is run to simulate the distribution system. Several runs must be made to achieve a calibrated model. Output, pressures, and flows from each run must be compared to actual field values. When modeled and actual values are within a desired tolerance, the model (or base data file) is said to be calibrated.

It is not sufficient to calibrate a model at only one demand condition. Compensating performance characteristics may lead to improper judgments regarding model calibration at a given condition. Calibration runs should be made at several demand conditions to verify the accuracy of the model over a range of flows. Adjustments are then made to nodal demands or pipe-friction factors, as judged most suitable in each instance, to fine tune the computer model.

From the calibration runs, the level of confidence in the model's capability to predict system performance is established. Further investigations may be necessary to resolve any major discrepancies. These discrepancies may occur from closed valves, incorrect pipe interconnections, incorrect pipe sizes, discrepancies in hydraulic coefficients, or other factors.

Costs of Calibration

The costs of calibrating a model vary according to the features of the distribution system. In general, the cost represents approximately 10 percent of the cost of the network analysis. This includes the field-data collection of physical data, such as pump calibration, pipe friction, and hydrant tests, and collection of operational data, such as short-duration measurement of actual system performance established using the base model.

Chapter 3

System Analysis

The principal objective of a network analysis is to determine the adequacy of the existing water distribution system facilities. Based on a computer model, network analysis can be used to identify the causes of deficiencies in the system and to develop the most cost-effective improvements. The analysis of water systems includes a number of interdependent variables, including pipes, pumps, and storage. System performance can be affected by altering various combinations of these components. Therefore, system analysis must be a comprehensive effort that addresses piping, storage, pumping, and operations. Table 3-1 summarizes the steps and activities involved in system analysis.

METHOD OF ANALYSIS

A distribution system may be analyzed using model-simulation programs or system-optimization programs. Simulation programs determine heads and flows at all nodes and pipes in the network. Optimization programs determine pipe sizes, pump heads, or pipe locations that optimize a particular objective, usually the lowest-cost solution.

This chapter describes system-analysis procedures for steady-state, model-simulation applications. The approach has been used successfully for system analysis in a wide variety of distribution systems. Neither dynamic simulation nor system optimization are addressed; however, most of the procedures described here are applicable to those approaches.

The procedures described in this chapter can be used in the hypothetical general case. Because general cases never strictly occur, it is important that the system analyst be knowledgeable in hydraulics and that sound engineering judgment be used in applying the procedures. System analysis is thus partly a science and partly an interactive trial-and-error art that incorporates the feel of the distribution system. This chapter is intended as a general guide, pointing out considerations of importance and providing some guidance on the scientific principles used in system analysis. The analyst should use this information as a base, and tailor the analysis to site-specific information.

Limiting Demand Conditions

Comprehensive steady-state analysis of a distribution system should be performed for a wide range of conditions. However, if a distribution system can operate satisfactorily under the most limiting and most severe demand conditions, it will operate satisfactorily for all conditions. For this reason, the demand conditions most limiting to the performance of system components should be established, and computer model runs should represent system operations at these most limiting demand conditions. In addition, operating conditions specific to the distribution system under study must be determined and input for each limiting demand condition. This is done during base model formulation (discussed in general in chapter 2).

Diurnal demands. The diurnal-demand curve developed as part of the system network data can be used to determine three limiting demand conditions. The curve should be established for a high-demand day and then extrapolated to represent the maximum-day demand. As shown on the idealized curve in Figure 3-1, four points on the curve are of specific importance. Points A and B represent two times during the day when the instantaneous demand equals the average maximum-day demand. Point D is the peak-hour demand, and point C is the minimum-hour demand, or the time when the maximum storage-replenishment rate occurs.

Ideally, a water distribution system should be operated at a constant water-supply rate with a consistent supply from water-source pumps at treatment plants and wells. Thus, on the day of maximum demand, it would be desirable to maintain a water-supply rate equal to the maximum-day demand rate. If supply is maintained at the constant maximum-day demand rate, the difference between the diurnal-demand

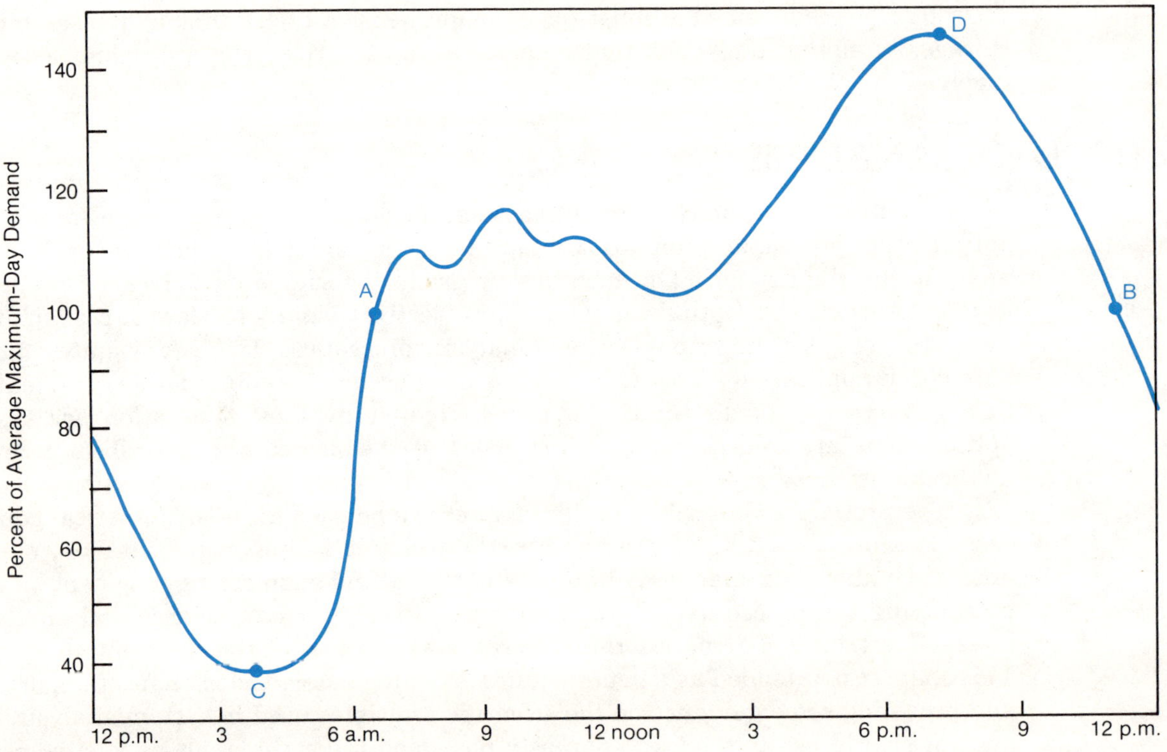

Figure 3-1 Idealized diurnal-demand curve.

curve and the maximum-day demand rate at any point in time would represent the flow into or out of storage facilities.

At the minimum-hour demand rate, represented by point C in Figure 3-1, the demand for storage replenishment is at its maximum. This is often a limiting condition that must be analyzed to determine whether the distribution system can provide this replenishment rate to the storage facilities.

At the peak-hour demand rate, represented by point D in Figure 3-1, flow out of the storage reservoirs is at its maximum rate. The storage reservoirs must provide outflow to meet the demand above the maximum-day demand rate. This is another limiting condition that must be evaluated to determine whether the distribution system can draw flow from storage and distribute it to meet the system demands at this rate.

Fire-flow demand. An important limiting demand condition that is not shown on the curve is fire-flow demand. According to the Insurance Services Office, fire-flow demands should be superimposed on the average demand of the maximum day. This occurs at points A and B on the curve in Figure 3-1. The most limiting of these points is B, because at this point storage facilities would have been used for equalization of demands and would be at a lower water level than at point A.

Peaking factors. Peaking factors are most-limiting demand conditions. Peaking factors are developed from the diurnal-demand curve, with maximum-day demand used as the base demand (Figure 3-2). The peak factors for the example diurnal-demand curve in Figures 3-1 and 3-2 are

peak-hour demand/maximum-day demand = 1.45

minimum-hour demand/maximum-day demand = 0.39

Typical ranges observed for these peak factors in distribution systems of various size are

peak-hour demand/maximum-day demand: 1.3–2.0

minimum-hour demand/maximum-day demand: 0.2–0.6

Additionally, a peak factor is generally developed for the ratio of maximum-day demand to average-day demand. This ratio has been observed to vary from 1.2 to 2.5.

Effect on system components. The various limiting demand conditions are most limiting to various components of the distribution system. In general, the relationship between limiting demand conditions and system-component performance is as follows:

The most-limiting demand conditions for system piping are maximum-day demand plus fire-flow demand, maximum storage-replenishment rate, and peak-hour demand.

The most-limiting demand conditions for system storage are peak-hour demand, and maximum-day demand plus fire-flow demand.

The most-limiting demand conditions for pumps are maximum-day demand, maximum-day demand plus fire-flow demand, and peak-hour demand.

Note that average-day demand is not included in the list of limiting conditions. Generally, average-day demand is a limiting condition only for pump selection, and it can be accommodated without individual model runs. Pumps are generally required to meet maximum-day demand, fire-flow demand, and/or peak-hour demand and are selected to have performance curves that allow operation through the full range of demands, including average-day demand.

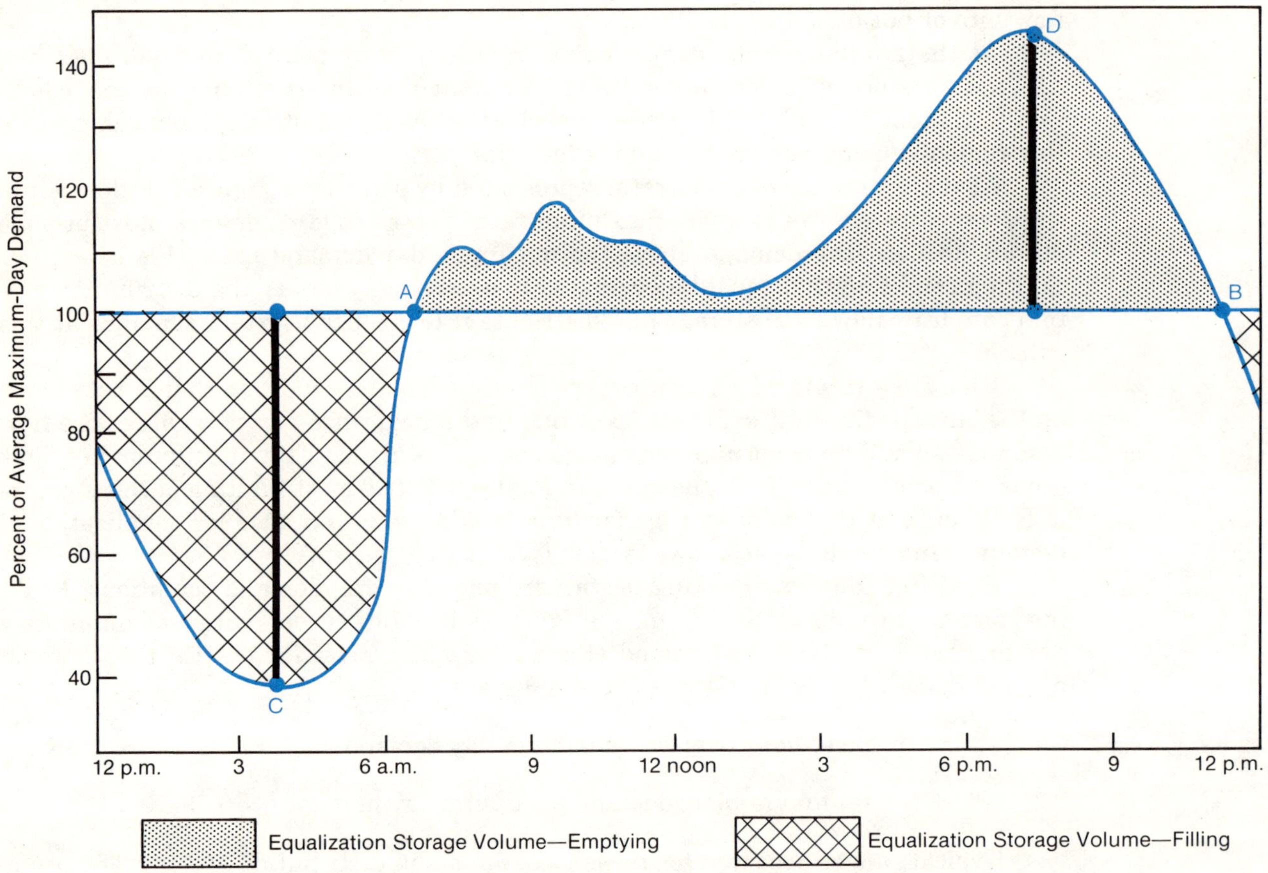

Figure 3-2 Maximum-day diurnal-demand curve.

Emergency Conditions and Reliability

The limiting conditions thus far discussed address the clearly identifiable periods of high summertime demand. The analysis of such situations is conducted under the assumption that all components of the system are operable. Infrequently, water systems are subject to emergency situations in which certain system components become inoperable. Examples of such situations include earthquake damage and well contamination. System analysis should address the degree to which the system can be relied on to function adequately when components are inoperable. The degree of reliability designed into the system is a policy decision that is based on a number of factors, including the probable frequency of the emergency condition, the cost of such a condition in terms of loss of service, and the cost of providing additional reliability.

Two emergency situations that occur frequently enough to warrant providing redundancy for better reliability are line breaks on critical mains and system-wide power outages. In anticipation of such situations, a policy judgment defining the acceptable level of service must be made. This level of service is defined in terms of a water-demand scenario, such as "the average-day demand should be met during a line break."

It should be remembered that even though the system may be capable of meeting the peak-hour demand, a critical line break or a 6-hour power outage may debilitate system performance to a greater degree. The method used to analyze system performance under emergency conditions depends on the engineering judgment of the analyst. The important objectives of such analyses are to identify potentially vulnerable portions of the system, investigate the results of system performance under emergency scenarios, and identify and figure the cost of alternatives to improve reliability. In this way, policy decision makers will have quantitative data with which to make decisions.

DESIGN CRITERIA AS ANALYSIS CONSIDERATIONS

Design criteria define system capabilities by specifying design features and performance requirements of system components. Thus, whether the objective of analysis is the design of a new system or improvements to an existing system, design criteria play important roles in defining the analysis process and as factors in network-solution calculations. In general, design criteria provide the standard against which system performance—both observed and predicted—is compared.

To understand the role of design criteria in system analysis, some review is necessary. A computer model is a mathematical tool that calculates the performance of a distribution system under various demand conditions. The model is used by first establishing the critical demand conditions to be evaluated and then reviewing the numerical output of the computer model. The output, or model-predicted performance, generally includes pressures and hydraulic grade lines at nodes, reservoir water level and flow rate, pipe velocities and flow rates, pipe head losses, and pump operating conditions.

System Deficiencies

To learn system deficiencies, the model-predicted performance must be compared to established design and operational standards. Deficiencies in a system are generally indicated by inadequate system pressures. In network analysis, model-predicted performance is evaluated to identify the system deficiencies that cause the inadequate pressures. Thus, the evaluation of model predictions must address the major system components, that is, piping, pumping, and storage, and also check the assumed system operating conditions compared to model predictions.

In analyzing model predictions, it must be recognized that a computer model either (1) calculates pressure, or hydraulic elevation, at each node when a demand is specified or (2) calculates demand or supply at each node when a hydraulic elevation is specified. For example, specifying a hydraulic elevation for a reservoir requires that the model calculate a replenishment rate (demand) or an outflow (negative demand). Conversely, specifying a replenishment rate or a supply at a reservoir requires that the model calculate a hydraulic elevation, or pressure.

Piping Systems

Deficiencies. In most applications, pipe segments are generally considered potentially deficient, or most-limiting segments, if they are predicted to have any of the following conditions:

- velocities greater than 5 ft/s,
- head losses greater than 10 ft/1000 ft, or
- large-diameter pipes (16 in. or greater) having head losses greater than 3 ft/1000 ft.

Velocities in pipe segments are acceptable up to about 10 ft/s. However, as velocities increase, pipe head losses increase exponentially and problems with water hammer develop. Generally, as velocities approach 5 ft/s, pipes become limiting factors in delivering water at acceptable pressures to the extremities of the system. When head losses in a pipe segment approach 10 ft/1000 ft, a substantial loss of pressure occurs in that length of pipe. Pipes with head losses exceeding 10 ft/1000 ft are generally a contributing factor to inadequate system pressures. Smaller head losses are necessary for large-diameter pipes, because they usually convey water long distances. In large-diameter pipes, head losses greater than 1 to 3 ft/1000 ft tend to be important because the head-loss accumulation becomes significant in long runs of pipe. However, none of these limits are hard and fast. The ultimate test of piping-system adequacy is the pressure at the point of delivery.

Design criteria. The design of new pipelines should address two basic features: sizing and routing. The size of a pipeline is established by first determining the maximum flow to be carried. The maximum flow is the projected limiting demand condition for that pipe segment. The demand condition that is most limiting for a pipe segment may be fire-flow demand, maximum storage-replenishment rate, or peak-hour demand. After the limiting demand condition has been established, the pipe size should be selected to limit velocities to 5 ft/s and head losses to less than 10 ft/1000 ft under that demand condition.

The second pipeline design feature to establish is routing. Potentially deficient line segments may be corrected by cleaning (pigging) or lining the pipe segment to increase its hydraulic capacity, replacing or paralleling the segment, paralleling adjacent segments, providing a new pipeline to route flows around the limiting segment, or adding pipeline to loop the deficient area (that is, supplying water from more than one direction). Selecting an improvement is done on a cost basis through a combination of engineering judgment and trial-and-error computer analysis. In general, looping should be used whenever possible. A looped system provides supply from two or more directions for large demands, such as fire flow, and provides backup or redundant supply paths in the event that a section of pipeline is taken out of service.

The location of new pipelines is determined, for the most part, by availability of right-of-way and construction easements and by using common sense. When locating new pipelines, consideration should be given to the projected growth characteristics of the service area, that is, in what direction growth will be, how far it will extend, and where major demand areas are expected to be. In many cases, extending water service to new customers outside the existing service area results in initial limited fire-flow capacities. Provisions should be made for looping these areas to supply the fire-flow demands.

Pumping Systems

Distribution systems may have both high-service (source) pumps and system booster pumps. General considerations, including the following, apply to evaluation of both. Pumps must be sized to meet the full range of system demands; thus, all of the limiting demand conditions, including average-day demand, may be conditions of importance for sizing the pumps.

Typically, constant-speed pumps are used for service and booster pumps, and system storage is used to equalize the pumping rate over the range of water demands. However, when storage space for equalization is limited, it may be desirable to use variable-speed pumps for booster or service pumps.

Generally, service and booster pumps are provided as multiple-pump installations. The pumping installation should be sized to provide maximum-day demand with any one of the pumps—preferably, the largest pump—out of service. It may be necessary to size pumps for maximum-day demand plus fire-flow demand. Individual

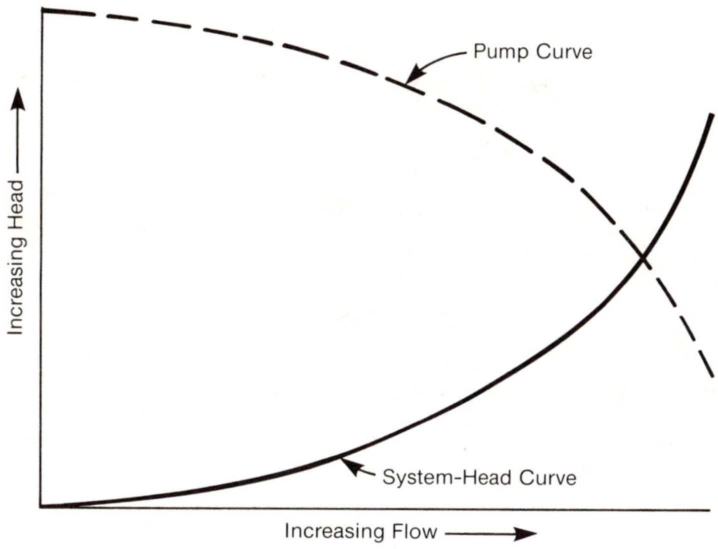

NOTE: The pump curve is a plot of the pump's capability to deliver flow at respective system head conditions. The system head curve is a plot of the system head that occurs at respective flows into the piping system.

Figure 3-3 Typical pump-characteristic curve and system-head curve.

pumps and combinations of pumps should be sized to meet the range of demands from average-day demand to maximum-day demand. In addition, the service pumps should be capable of providing maximum-day demand under peak-hour demand conditions, with the additional flow coming from storage.

A family of system head curves should be developed for service and booster pumps. These curves can be superimposed on characteristic curves for the pump combinations based on field-calibrated pump curves developed as part of base data file formulation. This family of curves should be based on a sequential operation of pumps in order according to field-established efficiencies at specific head conditions. Figure 3-3 shows a typical pump-characteristic curve and a system-head curve used to size pumps. The pump will operate where the two curves intersect, and the pump should be sized to operate at that point at its highest efficiency.

A system-head curve represents the required head for a range of flows. It is relatively easy to plot for a single pipe. It is calculated by adding the static head (elevation at discharge end minus elevation at source pump), the head loss through the pipe to the discharge end, and the head required at the discharge end of the pipe. However, the situation is more complex in a pipeline grid in which water is conveyed to many points through various routes rather than through a single pipeline to a single discharge point.

This situation is shown in Figure 3-4. On the left is the system-head curve and on the right is the family of hydraulic gradients, which slopes from the proposed pump to a point at which pressure must be at least 45 psi. The family of hydraulic gradients is the gradient between the pump and that critical node. In reality, the gradient is not a smooth curve as shown in Figure 3-4. As water is diverted to other demands, the head loss per pipe length changes along the route from the pump station to the critical node.

Figure 3-5 illustrates the results of a typical pump selection. The system-head curve is plotted and points on the curve at which flow into the system is expected to occur a high percentage of the time are noted. Depending on variations in system head versus demand, it may be possible to select pumps that operate efficiently over

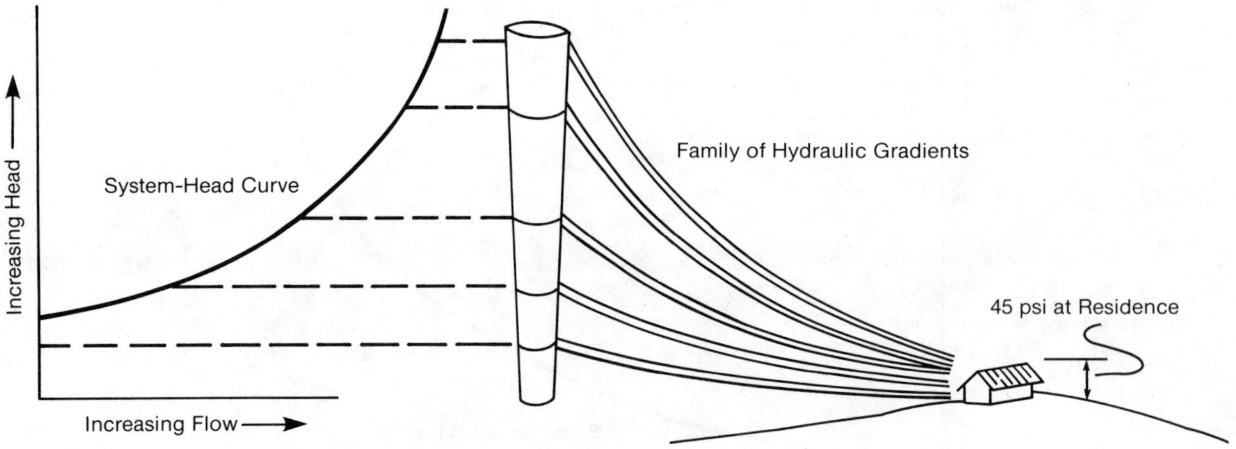

Figure 3-4 System-head curve for pipeline grid.

the range of demands or it may be necessary to select specific pumps for minimum-hour demand, average-day demand, and maximum-day demand.

In selecting pumps for a specific application, the efficiency of the pumps over the range of operating conditions must be evaluated. Figure 3-6 shows the typical efficiency range for a pump curve. The pump efficiency should be obtained from the specific pump curve for its points of operation on the family of curves. Pumps that have the maximum efficiency over the range of demands should be selected. In multiple-pump applications, the most efficient combinations of pumps should be determined and an operations plan established to maintain this sequential ordering.

Energy savings. Typically, pumping accounts for about 75 percent of the power consumed in a water system. Consequently, selecting the most efficient pumps for the range of demands and establishing an optimum sequential ordering of pumps have substantial influence on power consumption. Optimizing pumping efficiency has reduced power consumption in some water systems by 10–20 percent.

In addition to reduced consumption, careful scheduling can help reduce energy costs. For this purpose, the electrical rate structure should be analyzed. The rate structure may be based on not only monthly consumption, but also the demand factor (the maximum kilowatt demand for a short duration, such as 15 or 30 min), the loading factor (the ratio of maximum-demand rate divided by the average power usage), and ratchet clauses that extend these factors over many months (often 12 months). Pumping schedules should be established to minimize demand charges as much as possible. This can be accomplished by making the 24-h pumping schedules as uniform as possible, thereby minimizing the need for short-duration, high-power demands. This kind of scheduling may not apply to systems with variable energy costs.

Some electrical rate structures include time-of-use charges. In some cases, electrical utilities that do not have time-of-use rate structures are open to negotiating these rates for individual services. When time-of-use rate structures can be negotiated, the potential for savings by off-peak pumping, or peak shaving, should be evaluated. Peak shaving is accomplished by turning off one, several, or all pumps during peak electrical-cost periods and relying on storage to supply the water demand

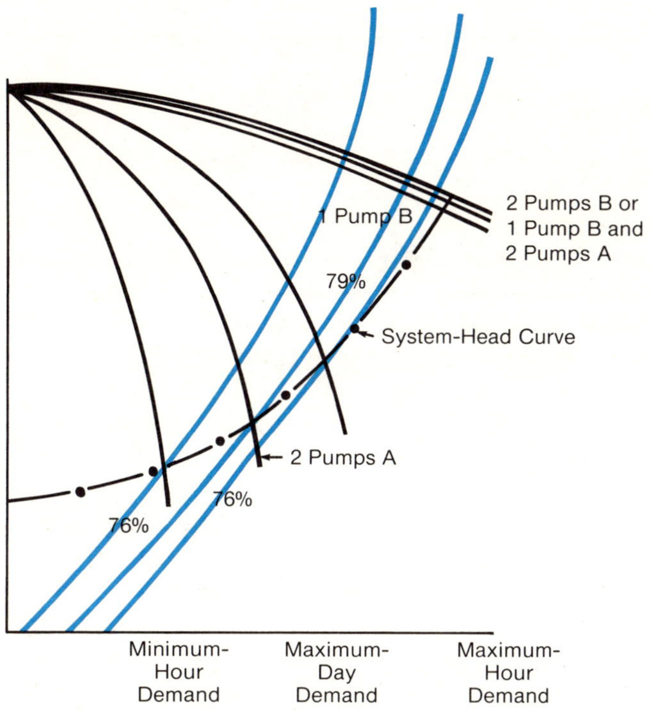

Figure 3-5 Typical pump-selection curves.

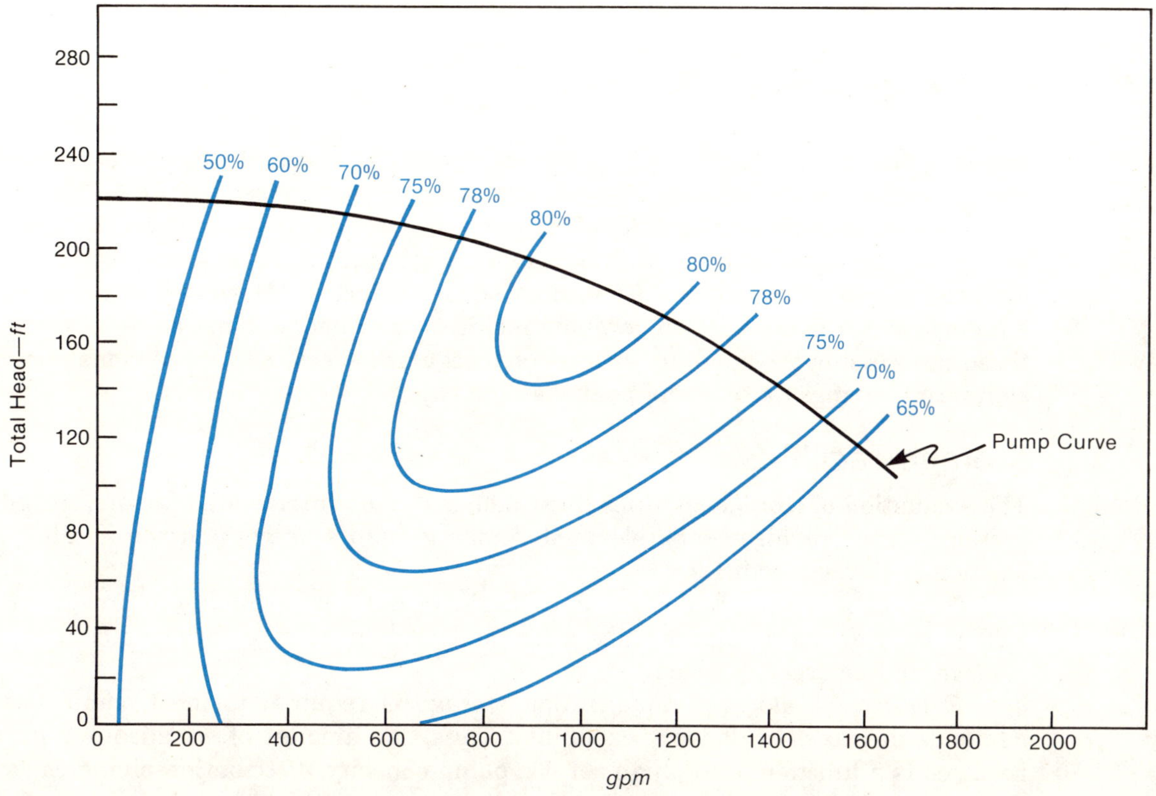

Figure 3-6 Typical efficiency range for pump-characteristic curve.

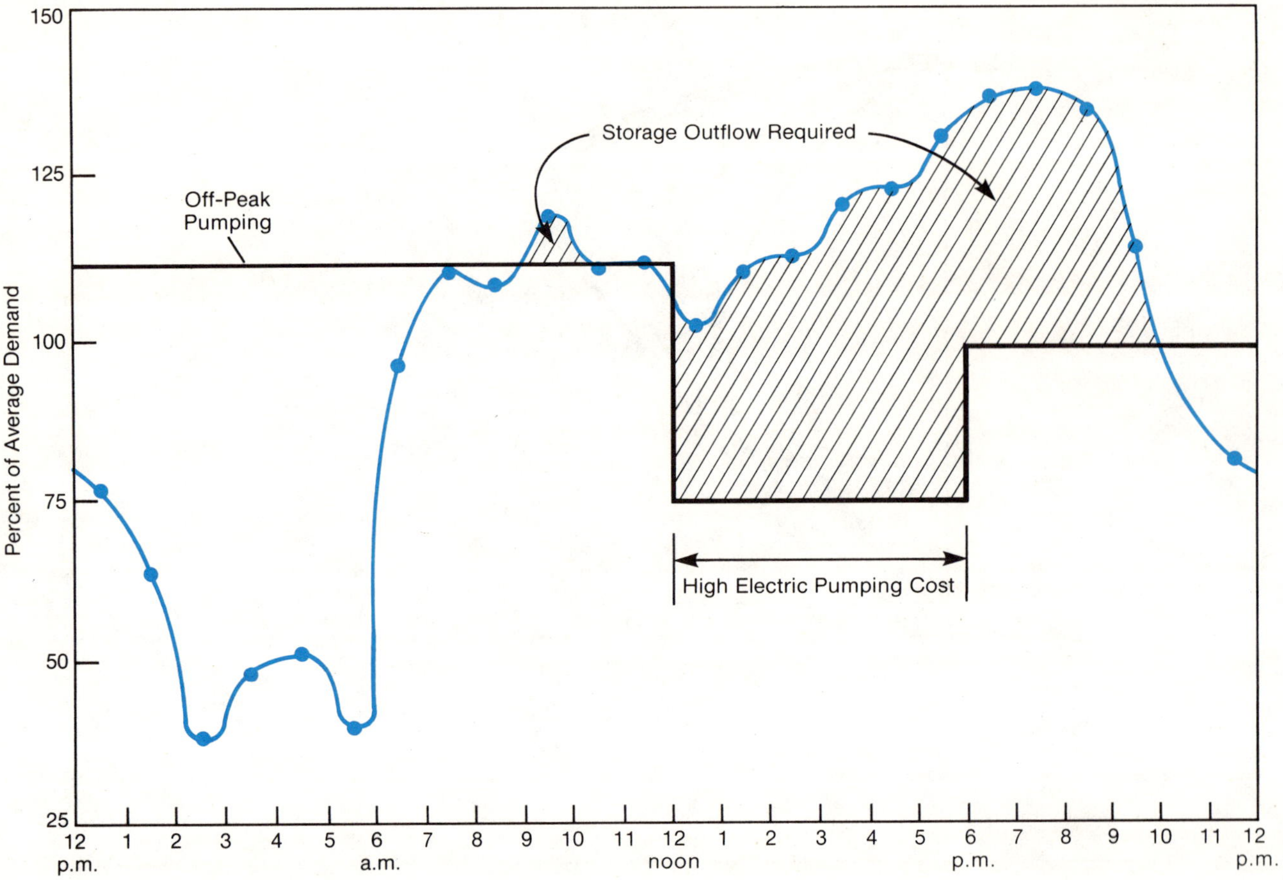

Figure 3-7 Alternate off-peak pumping schedules.

not supplied by pumping. This mode of operation generally requires additional storage and pumping capacity beyond normal requirements (see Figure 3-7).

As an alternative to turning off pumps, the use of auxiliary power or auxiliary drives may be used to shave peak demands. In general, using auxiliary sources only for peak shaving has not been found to be cost effective. However, in many cases auxiliary power is provided to pumping facilities to protect against power outages. In these cases, using the auxiliary sources on a regular basis to shave peaks may provide significant savings by reducing peak-demand charges.

Storage Facilities

The evaluation of storage facilities must address many interrelated factors, including system storage requirements, elevated versus ground storage, and the number and location of storage facilities.

System storage. Enough storage capacity must be available to provide clear-well storage for service-pump operation and system storage for equalization, fire reserve, and emergency needs.

Equalization storage. Equalization storage is required to meet water system demands in excess of delivery capability. Thus, the amount of equalization storage required is a function of the high-service pump capacity, distribution-piping capacity, and system demand characteristics. Equalization storage is generally less expensive to provide than increased capacities of high-service pumps and distribution piping beyond that required to meet maximum-day demand. Consequently, the pumping and piping systems are sized to carry maximum-day demand, and equalization storage is sized to carry demands in excess of the maximum-day demand up to peak-hour demand. Equalization storage generally makes up one half the total storage required and about 20–25 percent of the average-day demand; however, these percentages are not recommended as rule-of-thumb design standards.

Systems with excess pumping and piping capacity can decrease the amount of equalization storage by operating the system at a maximum pumping rate above maximum-day demand. As demands increase and the capacities of the pumping and piping systems are reached, it is generally most economical to draw from equalization storage. Using equalization storage may be more economical than using excess pumping and piping capacity. The benefits of operating the system at maximum-day demand rather than maximum pumping rate include cost savings through reduced power costs for pumping and through peak shaving. The network model should be used to evaluate various combinations of pumping, piping, and storage to select the most economical combination.

The idealized diurnal-demand curve is used to determine equalization storage volume. The typical maximum-day diurnal-demand curve in Figure 3-2 shows cross-hatched and dotted areas between the diurnal-demand curve and the maximum-day demand above and below the 100 percent maximum-day demand. These areas represent the required equalization storage volume. Storage facilities fill when demand is less than 100 percent maximum-day demand and empty when demand is greater than 100 percent maximum-day demand. Having adequate storage volume allows high-service pumps and treatment facilities to operate throughout the day at a constant rate equal to average demand. Equalization storage supplies all system demands above the average rate.

Fire storage. Typically, a water utility provides fire storage adequate for fire-flow requirements of the Insurance Services Office (ISO). The ISO determines fire-stream requirements throughout the distribution system's service area based on the characteristics of the individual buildings in the areas. Fire storage is then determined by ISO requirements for fire-stream durations, which are as follows:

Flow	Duration
Up to 2500 gpm	2 hours
3000 to 3500 gpm	3 hours
Greater than 3500 gpm	4 hours

Each storage facility should have enough fire storage to furnish fire-flow requirements within its area of influence. The area of influence is a function of area water-consumption demands, fire-flow demands, and distribution piping. Large fire-flow demands (in excess of 4000 gpm) may require more than one storage facility to overcome limitations in piping or other distribution features. Smaller fire-flow demands may require more than one facility, due to particular features of the distribution system. In large service areas, one reservoir may supply the entire fire flow for more than one fire stream. Reservoirs in higher-pressure zones may contribute to meeting fire-flow demands in a lower-pressure zone.

Fire-storage requirements should be predicted based on the expected number of storage facilities. Final fire-storage requirements should be established after total storage requirements are calculated and system storage locations are finalized.

Emergency storage. Emergency storage provides water during emergency situations, such as pipeline failures, major trunk main failures, equipment failures (such

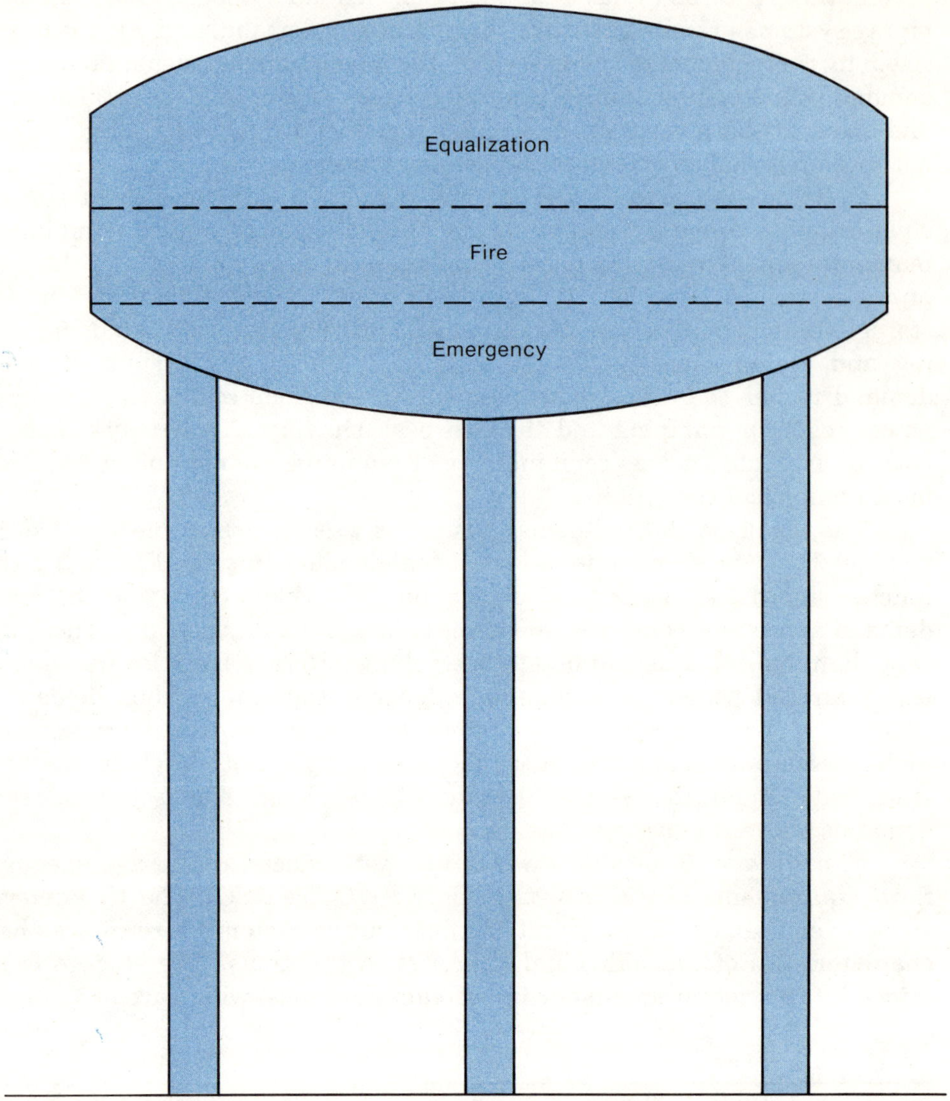

Figure 3-8 Schematic of system storage allocation.

as high-service or booster pumps), electrical-power outages, water treatment facility failures, raw-water supply contamination, or natural disasters. The amount of emergency storage included within a particular water distribution system is an owner option based on an assessment of risk and the desired degree of system dependability. An assessment must be made of the type and nature of emergency conditions, including their frequency, intensity, and duration. In general, a vulnerability analysis, such as described in *Emergency Planning for Water Utility Management,** or a more limited vulnerability assessment should be used to determine emergency-storage requirements.

**Emergency Planning for Water Utility Management*, AWWA Manual M19, AWWA, Denver, Colo. (1984).

Unlike equalization and fire storage, which should be available at all system storage sites, emergency storage may be included at only one or a few of the storage sites.

Storage allocation. In evaluating system storage and developing operational plans for its use, the allocation of storage for equalization, fire, and emergency must be addressed. In doing so, it may help to think of storage schematically, in terms of its position in the storage tank. In general, system storage should be allocated as shown in Figure 3-8. Storage for equalization should occupy the top portion of the tank. Because storage for equalization amounts to about one half of the total storage volume, storage tanks would vary from full to depletion-of-equalization volume (or half full) as demand varies throughout the day. Fire storage is positioned next, "under" equalization storage, and emergency storage is at the bottom of the tank. The importance of designating levels for types of storage is to ensure that the storage volume has the correct hydraulic grade line for the intended purpose. For example, fire storage may be required at a time when the equalization storage has been depleted; this schematic allocation of storage allows for the resulting change in the hydraulic grade line. System analysis should ensure that fire storage is available at the hydraulic grade line equal to the schematic "bottom" of fire storage.

Elevated versus ground storage. The evaluation of elevated and ground storage should address high-level ground storage, elevated storage, and ground storage with booster pumping (see Figure 3-9). If there is high ground in or near the service area, ground storage connected directly to the system may be used. However, for most distribution systems, the use of elevated tanks, ground storage with booster pump, or a combination of both must be evaluated.

There are advantages to both elevated storage and ground storage. Elevated tanks store water at the hydraulic elevation maintained by the system. Booster pumps and controls are not required, which results in simpler operation. Elevated storage is more reliable than ground storage because elevated storage provides water

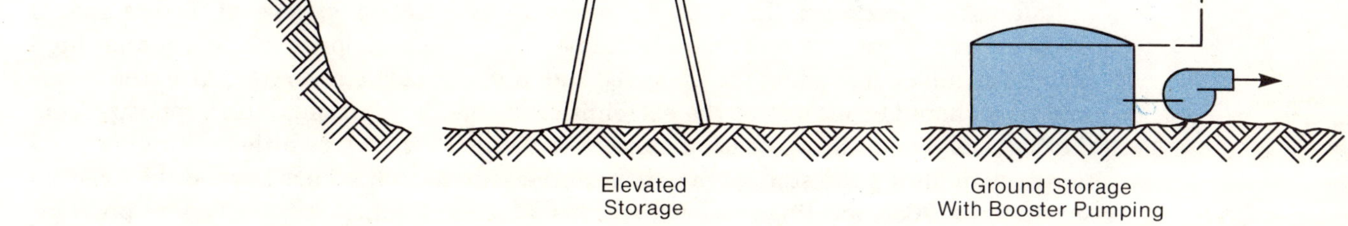

Figure 3-9 Typical storage arrangements.

even during a power failure. Finally, elevated storage provides stable system pressures.

Construction costs. A major advantage of ground storage is lower construction costs. Depending on system features, the initial construction cost for elevated storage may be 2–4 times more than ground-storage construction costs per unit of storage. Further, ground storage may be built to hold 5 mil gal or more, whereas the practical size limit of elevated storage is about 3 mil gal. Established ground-storage facilities can be adapted to changing system hydraulics economically by replacing or modifying pumps and controls. In contrast, elevated tanks must operate at a specific hydraulic elevation. Frequently, as water systems grow and new customers are added farther away from elevated tanks, operating pressures are increased and tanks stay full or nearly full with no storage available for equalization. Finally, ground-storage facilities can be regulated to operate in response to system demands. In contrast, elevated storage may, in some cases, deplete too soon or fill during demand periods.

Operation and maintenance costs. Steel elevated tanks generally have higher costs for painting and corrosion protection than ground storage. Power costs may vary according to the conditions under which the facility is operated. Ground storage has power costs for booster-pump facilities; however, elevated storage maintains higher pressure at average operating conditions, which results in higher pumping head requirements and higher power costs per gallon pumped.

Number and location of storage facilities. A number of factors must be considered when determining the number and location of storage facilities. These include size of reservoirs; the location of other system components, such as major trunk lines and high-service pumps; hydraulic elevation; and zone of coverage for fire flow.

Number of reservoirs. The number of reservoirs required for a system should be established using trial-and-error analysis. The analysis should develop a cost-effective number of storage reservoirs, taking into consideration the impact in terms of cost, on piping, pumping, and storage. Engineering judgment will limit the number of reservoirs to evaluate.

A key factor to establish is the limiting size of each reservoir. The limiting size depends on whether ground or elevated storage is to be used. In general, reservoirs should be as large as practical, but small enough to be filled in a reasonable amount of time. For example, a water distribution system that has four major service areas with high demands could consider two options. One reservoir could be put in each of the four areas or regional reservoirs could be placed in two areas. A calculation of the volume requirements of each reservoir and a check of piping and pumping requirements for both options will indicate whether four smaller reservoirs may be less costly than two larger reservoirs due to piping and pumping requirements for the larger regional reservoirs.

Reservoir locations. In selecting reservoir site locations, several factors should be considered. First, the reservoir should be located on a major trunk line so that high flow rates can be adequately transferred into and out of the reservoir. Also, the reservoir sites should be located at the extremities of the distribution system opposite from the high-service pumps. Locating storage facilities at the extremities of the distribution system allows the storage facility to serve growth that occurs beyond the existing service area. Also, locating storage facilities at the extremities of the system provides a more equal pressure distribution throughout the whole service area. Figure 3-10 illustrates the effect of the reservoir location on the system hydraulic grade line.

For effective fire flows, the storage site should be located to provide coverage over a service area, while avoiding unnecessary overlap with other storage sites. It is beneficial to locate a storage site as close as possible to areas requiring high fire

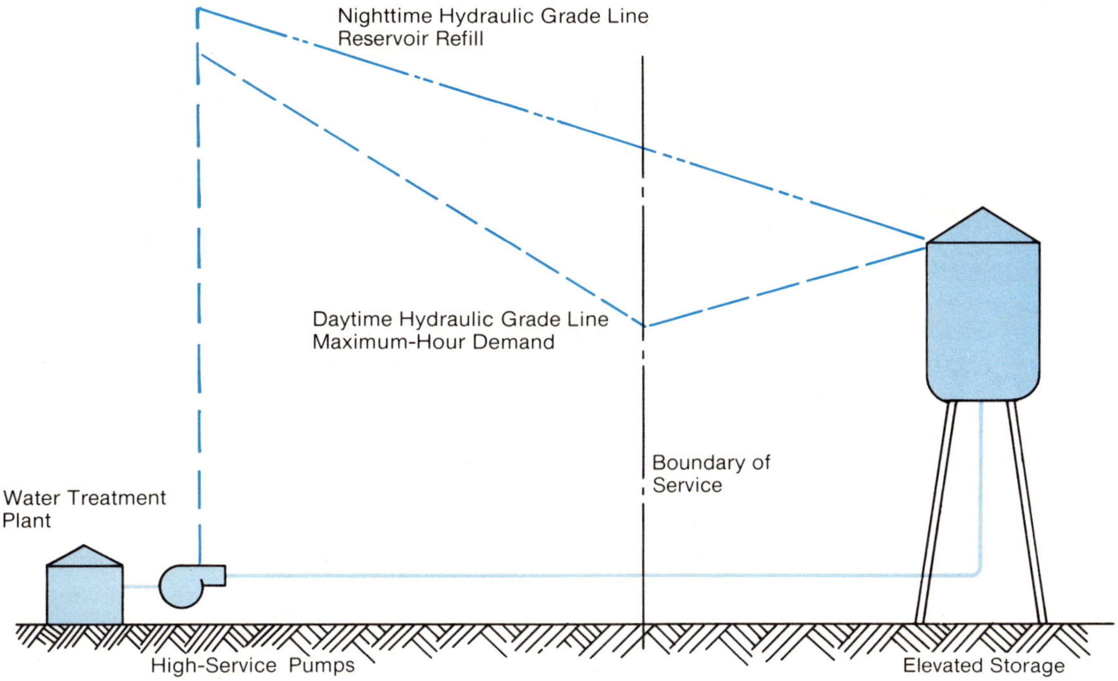

NOTE: The system hydraulic grade line varies significantly according to the time of day.

Figure 3-10 Impact of elevated storage location on system hydraulic grade line.

flows. For effective emergency flows, the storage site should be located as it was for fire-flow requirements–that is, close to major demands—rather than spread out through the water system.

Hydraulic elevation. The overflow elevation of a reservoir must be set correctly to assure its proper operation. The elevation must be low enough so that the reservoir may be filled by the high-service pumps, but high enough to supply adequate pressure through the range of demands (especially peak-hour demand) throughout the system. When several reservoirs are on a single distribution system, they must be arranged to ensure that reservoir levels are balanced and the storage is used equally. Ground storage facilities should be provided with booster-pump facilities sized to meet the range of demands (especially peak-hour demand) at the appropriate system pressures.

ANALYSIS PROCESS

In network analysis, a field-calibrated model is used to evaluate a distribution system for both existing and projected limiting demand conditions. Usually, the computer model is run for maximum-day demand, maximum storage-replenishment rate, maximum-day demand plus fire-flow demand, and peak-hour demand, and deficiencies in system performance are evident as unacceptably low or high pressures or as inconsistent operating conditions.

Before beginning analysis, the analyst should prepare an outline or index of model runs that describes the situation simulated. Each model run should be numbered or named, and a copy of each run's printout should be kept in a binder. This helps to organize the large amount of data that is generated during multiple model runs.

It is advisable to make only one or two model runs at a time. Each model run should be studied individually and the results recorded before performing subsequent runs. The analyst should make notes on the printouts so that the evaluations will be readily understood by anyone referring to them in the future.

Pressure

Design-pressure criteria may vary with the particular features and conditions of individual distribution systems. In general, however, the desired range of system pressures is between 30 psi and 90 psi through the range of system demands, including average-day demand, maximum-day demand, maximum storage-replenishment rate, and peak-hour demand. Under fire-flow demand conditions, it is generally acceptable to maintain system pressures of 20 psi. Model-predicted pressures outside the desired range of system pressures indicate the extent of distribution system limitations in meeting demands. These limitations may be caused by deficiencies in piping, storage, pumping, or a combination of these deficiencies.

Negative pressures. Depending on the extent of distribution system deficiencies, the network model may predict negative pressures. These predicted negative pressures generally do not actually occur in the system. They appear in model predictions because the mathematical model is forced to supply the full amount of all nodal demands. The significance of predicted negative pressures is that the system demands cannot be met. In some cases, negative pressures may actually occur in limited areas of the system, for example, in high-elevation areas. However, in general, a deficient distribution system simply fails to meet demands—that is, system pressures decrease as the demands increase, and a lesser demand that can be sustained by low system pressures is reached.

Model Initialization

Model initialization involves selecting or setting and inputting distribution system variables for specific model runs (demand conditions). The system variables are the conditions at the pumps, pressure-regulating valves, control valves, and storage facilities.

There are several ways to initialize a model. Features and limitations of existing facilities, such as storage capacity, must be recognized and included in the computer model to allow for analysis of the actual system. This section describes two approaches to model initialization for limiting demand factors. The first approach assumes that adequate system storage is available for equalization. The second approach is for systems with less than the desired amount of equalization storage.

Initializing for systems with equalization storage. Systems with equalization storage use pumps and other system components differently than systems without such storage. Therefore, the method used for initialization depends, in part, on storage. The method for systems with adequate storage for equalization assumes that service pumps operate at the maximum-day demand rate with storage contributing flow and refilling as demand conditions change.

Maximum-day demand. Maximum-day demand is often a limiting condition for service and booster pumps. The maximum-day diurnal-demand curve in Figure 3-2 shows that maximum-day demand occurs at points A and B. On this curve, point A is a more limiting condition for elevated storage than point B. At point A, storage facilities are full and the requirement to sustain system pressure is the highest; equalization storage is depleted at point B. Storage facilities do not contribute to the system during maximum-day demand; thus, they should be represented in the system model as nodes with no demands or hydraulic elevation, unless unique circumstances dictate that a hydraulic grade line be established. All water should come to the system from the water-source pumps. For this reason, source pumps should be initialized so that their combined production is adequate to provide maximum-day demand. When all storage facilities are entered as nodes, the source pumps, during maximum-day demand, will be required to meet the full system demand; thus, the pump curves should be entered. Finally, the distribution piping may be a limiting condition because water is obtained from only the water source and is conveyed through the entire pipe grid.

In reviewing the model runs for the maximum-day demand condition, the storage nodes should be examined for predicted hydraulic grade lines. The service pumps and piping system should be capable of delivering water to the reservoirs at a pressure level equal to the hydraulic elevation that is equivalent to the top of the storage tanks. If the predicted hydraulic grade line is below the top of the storage tank, storage would contribute water before maximum-day demand is reached, indicating that storage would not be filled before maximum-day demand or that it would be depleted too soon.

Maximum-day demand plus fire-flow demand. Maximum-day demand plus fire-flow demand is a limiting condition for service pumps and the piping system. It is also a limiting condition for storage facilities and booster pumps when fire-flow demand occurs within the areas they serve. In large systems, the fire-flow demand condition is generally less critical than the peak-hour demand condition for the overall system; however, in small systems and within the service area affected by fire-flow demand in large systems, the localized impacts on distribution piping may be more limiting.

When initializing the model for maximum-day demand plus fire-flow demand, service pumps should be initialized in combinations adequate to provide the maximum-day demand. The additional fire-flow demand should be provided by storage facilities and booster pumps. The limiting fire-flow demand condition is at point B on the diurnal-demand curve, as in Figure 3-2. Ideally, at this point equalization storage throughout the system is depleted. Initialization for storage facilities outside the fire-flow influence area should be at the hydraulic grade line equivalent to the tank elevation when equalization storage is depleted. The storage remaining in these facilities is the fire-flow storage and emergency storage. Generally, these storage tanks would be approximately half full at this point. Initialization for storage facilities within the fire-flow influence area should be as contributing reservoirs. The hydraulic grade line for these storage reservoirs should be equivalent to the tank elevation when both equalization storage and fire storage are depleted; the only storage remaining is emergency storage. Generally, these storage tanks would be almost empty at this point. This configuration is essential to ensure that all fire-flow storage under this limiting condition is available to the system.

A more conservative approach to model initialization for the maximum-day demand-plus-fire-flow demand condition involves initializing storage facilities outside the fire-flow influence area as for the maximum-day demand condition, that is, as nodes with no demands or hydraulic elevation. As before, storage facilities within the fire-flow influence area are initialized as reservoirs at a hydraulic grade line equivalent to equalization and fire-storage depletion. In this approach, the service pumps are initialized at a pumping rate (negative demand) equivalent to maximum-day demand. By setting this delivery rate, the network model is forced to draw fire-flow demand from the contributing storage reservoir. This configuration forces storage facilities to provide fire-flow storage for their areas of influence, which is the desired operating condition. This very conservative design approach prohibits adjacent storage facilities from assisting in meeting fire-flow demand conditions outside their areas of influence. In reviewing model runs for this conservative approach, the storage nodes outside the fire-flow influence area should be examined for predicted hydraulic grade lines compared to expected hydraulic elevations. The service pump and piping system should be capable of delivering water at a pressure equal to the desired hydraulic grade line.

Maximum storage replenishment. The maximum storage-replenishment rate rate is a limiting condition for the piping system. It may also be a limiting condition for service and booster pumps, depending on the system pressures at maximum storage-replenishment rate versus pressures at maximum-day demand conditions. Storage facilities exert large, concentrated demands that must be met by conveying water through a limited pipe path. When storage-replenishment demand is at its maximum, major trunk mains often experience their most limiting condition.

Again, service pumps should be initialized in a combination adequate to provide the maximum-day demand. Although maximum storage-replenishment rate is demanded when storage facilities are not yet approaching full (see Figure 3-2, point C), storage facilities should be initialized at their overflow elevations. This provides some margin in pipe and pump sizing to assure these systems are capable of completely filling the tanks. It also provides the operator some safety margin if the system falls behind in storage. All storage facilities are initialized as reservoirs with hydraulic elevations at the top of the tank or with demands equivalent to the maximum storage-replenishment rate. This replenishment demand is the difference between the maximum storage-replenishment rate at point C on the idealized demand curve and the maximum-day demand allocated among storage facilities based on the ratio of individual facility equalization storage to total equalization storage.

Peak-hour demand. Peak-hour demand is the condition that establishes equalization storage requirements. It is generally the most limiting condition for the system storage and booster pumps. It may also be a limiting demand condition for system piping; however, maximum-day demand plus fire-flow demand is generally more limiting for piping. Peak-hour demand from storage, as shown in the diurnal-demand curve in Figure 3-2, is the difference between peak-hour demand at point D and maximum-day demand. Thus, storage facilities are initialized to provide an outflow (negative demand) allocated among storage facilities based on the ratio of individual facility equalization storage to total equalization storage. In reviewing model runs, the predicted hydraulic elevations at each storage facility should be compared to the expected hydraulic grade lines. The expected hydraulic elevation at each storage facility is equivalent to the storage-tank level at peak-hour demand, that is, at point D on the diurnal-demand curve in Figure 3-2. This elevation can be established by determining the volume of equalization storage to be used prior to point D and then calculating the water level in each tank after this volume draws down. If the model-predicted hydraulic elevation is significantly lower than expected, the storage would be drawn down or depleted before the peak-hour demand condition is over. If the model-predicted hydraulic elevation is significantly higher than expected, storage would not be available as needed during peak conditions.

An alternate approach to initializing peak-hour demand is to initialize the storage facilities as reservoirs at the desired hydraulic elevations and allow the model to predict the outflow. The predicted storage outflows are compared to the desired outflows. In some instances, system storage facilities have been observed to be filling rather than providing water to the system at peak-hour demand. This is obviously

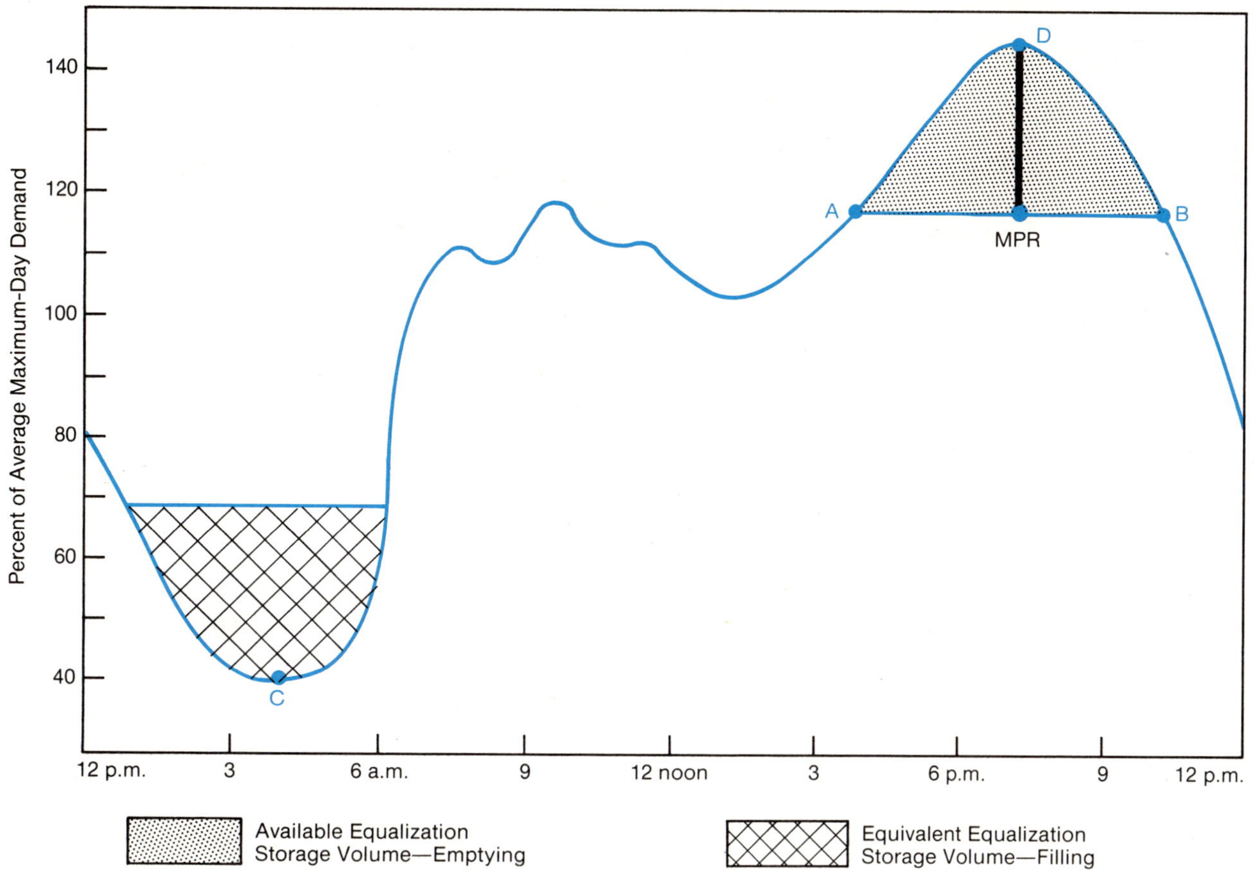

Figure 3-11 Maximum-day diurnal-demand curve with maximum pumping rate.

undesirable, and system controls or improvements should be developed to eliminate this condition. For modeling purposes, reservoirs that fill during peak-hour demand may be initialized with a desired hydraulic elevation. The computer then predicts an inflow to these reservoirs.

Initializing for systems with inadequate storage. In some systems, the existing available storage may not be adequate to allow the service pumps' continuous operation at the maximum-day demand rate. The distribution system may accommodate the storage limitations in several ways. One common method is to operate the high-service pumps at a maximum pumping rate greater than maximum-day demand. The maximum pumping rate required to accommodate storage limitations can be determined by analyzing the existing storage volume with respect to the existing diurnal-demand variations.

Figure 3-11 is a typical maximum-day diurnal-demand curve for a system with limited storage. The dotted area represents the limited existing storage volume. The horizontal line below this area, connecting points A and B, represents the required maximum pumping rate. When initializing a model based on this curve, it is necessary to specify operational conditions at the limiting demands that recognize the required maximum pumping rate. Consequently, for the four limiting demand conditions, the service pumps would be initialized at maximum pumping rate rather than the maximum-day demand. Storage outflows and replenishment rates would be calculated from the maximum pumping rate.

Table 3-1 Model Initialization for Typical Limiting Conditions

Operational/ Demand Condition	Service Pumps	Booster Pumps	Storage
Maximum-Day Demand or Maximum Pumping Rate	Enter pump curves or set equal to maximum-day demand or maximum pumping rate as negative demand.	From storage: Initialize as out-of-service, as nodes with no demands. In-line boosters: Initialize as in-service, enter pump curves.	Initialize as nodes with no demand.
Maximum Storage-Replenishment Rate	Initialize as negative demand equal to maximum-day demand or maximum pumping rate.	From storage: Initialize as out-of-service, as nodes with no demands. In-line boosters: Initialize as in-service, enter pump curves.	Initialize as reservoirs with water surface at overflow elevation or with negative demands equal to maximum replenishment rate.
Maximum-Day Demand plus Fire-Flow Demand	Initialize as negative demand equal to maximum-day demand or maximum pumping rate.	From storage: Initialize as out-of-service, as nodes with no demands. In-line boosters: Initialize as in-service, enter pump curves.	Outside area of influence: Initialize as reservoirs with a water level at which equalization storage is depleted. In area of influence: Initialize as reservoirs with water level equal to emergency storage level.
Peak-Hour Demand	Initialize as negative demand equal to maximum-day demand or maximum pumping rate.	From storage: Initialize as nodes with negative demand equal to storage outflow, or as in-service by entering pump curves. In-line boosters: Initialize as in-service, enter pump curves.	Initialize as nodes with negative demands equal to storage outflow. Or initialize as reservoirs with water level at the desired hydraulic elevation.

Summary. A summary of the suggested model initializations for typical limiting operational and demand conditions are listed in Table 3-1. All model initializations require that a hydraulic elevation be input for at least one location. This is necessary to establish a hydraulic elevation for the analysis.

Assessing Model Runs

Model runs should be assessed by evaluating system components, including piping, storage, and pumping.

Piping. The primary evaluation of system piping is made under peak-hour demand conditions. The evaluation should also include maximum-day demand plus fire-flow demand, as well as maximum storage-replenishment rate, conditions. Depending on the features of a particular distribution system, the maximum storage-replenishment rate condition may be most limiting for the major trunk mains. During maximum storage replenishment the system meets the high demands concentrated at storage facilities by conveying water through a limited pipe path, or the major trunk mains.

In assessing model runs in regards to piping, the predicted conditions for each individual pipe segment are evaluated in order to assess the performance of the pipe segment. The head loss per pipe segment should be examined, but whether this is an indication of potential deficiency depends on the length of the pipe segment. The velocity predicted for each pipe segment and the head loss per unit length (1000 ft) should be examined. The pipe segments identified as potentially deficient are those most limiting to delivery capacity and the largest energy consumers. This approach to identifying potentially deficient pipe segments is finalized through model runs of proposed improvements.

Storage and associated booster pumps. An overall assessment of system storage is performed as part of the system analysis. To evaluate the performance of the existing storage and booster-pump facilities, computer model runs are analyzed at maximum-day demand plus fire-flow demand, as well as peak-hour demand, conditions. The adequacy of storage facilities and booster pumps is assessed by reviewing predicted hydraulic elevations and predicted system flows, as well as pressures and pipe velocities throughout the piping network.

High-service and booster pumps. The evaluation of high-service and booster pumps is made primarily at maximum-day demand. Additionally, maximum-day demand plus fire-flow demand, as well as peak-hour demand, may be limiting conditions. Depending on the features of the particular distribution system, maximum storage-replenishment rate may also be a limiting condition. Predicted system pressures and flows are compared to pump characteristics established in the family of curves. From this comparison, the adequacy of high-service pumps is assessed.

Sensitivity Analysis

After the model has been calibrated to establish the limits of accuracy, or level of confidence, in model predictions, a sensitivity analysis should be performed. Although sensitivity analysis is not essential, it provides greater confidence in model predictions and a better understanding of the model's response to variations in key parameters. The sensitivity analysis should address variations in the following system features:

- increase in pipe roughness (typically 10–20 percent),
- decrease in pipe roughness (typically 10–20 percent),
- increase in limiting demand conditions (typically 15–25 percent), and
- location of new large demands at several locations through the system.

It is also desirable to assess the model's sensitivity to storage-tank levels and pumping options. In many cases, storage and pumping have much more influence on system performance than pipe roughness.

Depending on the results of the sensitivity analysis, it may be desirable to perform similar sensitivity evaluations in the model analysis of limiting demand conditions or in the development of improvements.

DEVELOPING SYSTEM IMPROVEMENTS

A principal objective of network analysis is to develop and evaluate potential system improvements. A comprehensive plan for system improvements that addresses all deficiencies and needs identified in the system analysis should be developed. The plan must include some flexibility to provide for real-world limitations in system

Develop power-management program.

- An evaluation of energy requirements and development of a power-management program should be performed concurrently with development of system improvements.
- The power-management program should address alternate system operations, load factor improvements, reducing pumping rates, and reducing operating pressures.

Develop distribution system improvements.

- Based on system deficiencies, alternate system improvements should be evaluated for the distribution system.
- The network model should be used to evaluate proposed improvements.
- Develop the most economical or cost-effective improvements, including distribution piping, storage, service pumps, and booster-pump facilities.
- Evaluation of proposed improvements should take into consideration not only capital and operation and maintenance costs but also noncost factors, such as reliability, flexibility, and ease of operations.

Develop an operational improvement program (OIP) and system control and data acquisition (SCADA) program. (These two programs should be developed concurrently.)

- The OIP includes an analysis of operation and control method to determine the most economical and energy-efficient system or procedure.
- Based on the recommended OIP and power-management program and the recommended distribution system improvements, develop a SCADA program.

Figure 3-12 Developing system improvements.

performance. The use of engineering judgment in conservatively applying engineering design standards provides this flexibility.

Three principal tasks are generally accomplished in developing a plan for system improvements (see Figure 3-12). They include energy, or power, management, the development of a number of suggested options for improvement, and development of programs for operational improvements (OIP) and system control and data acquisition (SCADA).

Power Management

The potential energy savings from power-management improvements can be as much as 10 or 20 percent of the total energy cost. For this reason, evaluating energy requirements and developing a power-management program are key steps in the development of distribution system improvements. The following items should be considered in the process:

- power consumption by system component;
- alternate system operations to reduce energy consumption, such as operating pumps in sequential order according to efficiency;
- potential savings through load-factor improvements, such as peak shaving by stand-by power;

- reducing maximum pumping rates to at least the average maximum-day demand rate by providing necessary piping, pumping, or storage improvements; and
- reducing system operating pressures at the water treatment plant by providing necessary piping, pumping, or storage improvements.

Options for Improvement

Based on system limitations and deficiencies, improvements to the system should be developed and evaluated. These potential improvements should be comprehensive, covering piping, pumping, and storage. Proposed improvements that are usually analyzed with the computer model include high-service pump modifications and additions, system storage additions, booster-pump modifications and additions, regulating- and control-valve modifications, and distribution-piping improvements. Development of improvements is an iterative, trial-and-error process. The initial development of improvement alternatives should be based on engineering judgment; the alternatives should then be evaluated using the computer model. Cost analyses should be performed to select the most cost-effective overall improvements. (As mentioned at the beginning of this chapter, system optimization programs can be used to develop improvements to meet cost objectives. Optimization is discussed in chapter 4.)

Alternative-improvements analysis should evaluate power-management alternatives and consider storage versus pumping options; piping versus pumping options; storage versus piping options; and the formation, deletion, or revision of pressure zones. The alternative-improvements analysis generally begins with an initial analysis based on maximum-day demand and peak-hour demand model runs. The final development of improvements incorporates all four limiting conditions: maximum-day demand, maximum storage-replenishment rate, maximum-day demand plus fire-flow demand, and peak-hour demand.

For systems with large anticipated growth, it is important to coordinate short-term and long-term improvements. One approach is to first establish a skeletal plan for the anticipated demand situation at the end of the planning period, and then "fill in" the short-term improvements. In doing so, the engineer should be aware of the uncertainty of the far-distant future-demand pattern. With this warning in mind, the establishment of a long-term plan establishes a framework for supply sources, major trunk mains, and storage within which shorter-term, more detailed improvements can be made. By considering the long-term first, and the short-term second, the detailed short-term improvements can be better coordinated within the long-term plan.

Operational Improvement Program and System Control and Data Acquisition

An operational improvement program (OIP) and system control and data acquisition (SCADA) program may be developed concurrently. The OIP incorporates the power-management program and should be consistent with recommended distribution system improvements. The OIP should include an analysis of alternate operation and control methodologies to determine the most economical and energy-efficient systems and procedures. This analysis should consider various combinations of manual and automated control techniques to tailor a control system or procedure for the given distribution system and operating staff.

Development of the SCADA program should be based on the recommended OIP, the power-management program, and recommended distribution system improvements. Distribution system monitoring should complement the system's OIP and should cover key distribution components or conditions, including tank storage depths, system pressures at key locations, system flows at key locations, service-pump operations, booster-pump operations, control-valve operations, and development of a distribution system control center. Monitoring techniques should be customized to the individual system; the planner should consider alternate methods, such as continuous recorders, telemetry, instantaneous gauges without telemetry, manual controls, automated controls, computerized controls, or various combinations of these methods. The SCADA program may also provide for remote operation of high-service pumps, booster pumps, and control valves. A central processing unit may be useful for automated system control and data storage.

CONTINUING USE OF THE MODEL

A utility that invests the time, effort, and cost in performing a computer-assisted analysis of its water distribution system should consider maintaining the model for continued use. Maintaining the model for continued use offers many advantages. Among other things, it

- allows for consideration of variations in system operational conditions, such as supply rate and head, storage water levels, and booster pump operations;
- provides for analysis of various operating and control systems;
- provides a planning tool to analyze the system's ability to serve additional large demands, such as industries, major subdivisions, annexation areas, or adjacent utilities;
- allows analysis of future conditions;
- provides for the development of the optimum or most cost-effective overall improvement plan by allowing evaluation of alternatives, such as piping improvements versus expanding pumping capacity, or storage improvements versus expanding piping or pumping capacity; and
- provides for performing vulnerability analysis, such as the effects of removing an individual facility component from service.

There are many methods of maintaining and using the model. These may range from hiring an outside consultant to provide the service, to the utility itself maintaining the model. If the utility elects to maintain the model, the utility staff should be properly trained in hydraulic analysis and computer modeling.

Chapter 4

Optimization of Design and Operation

Optimization is sometimes used rather loosely to mean any analysis aimed at improving the way something is done. This chapter discusses optimization in a more limited sense—that is, the formal mathematical optimization of the design and operation of a water distribution system. Optimization in this sense means achieving the best possible solution for a particular objective, which is usually lowest cost.

There is less practical experience with optimization of distribution networks than with network analysis and simulation. Optimization methods are taking a considerable amount of time to move from research, through development and testing, into practice. So, while other chapters in this manual describe what is being done in engineering practice, this chapter deals with methods that exist, but have been applied only to a limited number of cases. Methods that show the greatest promise for practical use are described in detail in this chapter, while other methods are mentioned for the sake of completeness.

ROLE OF NETWORK ANALYSIS IN OPTIMIZATION

Optimization is a mathematical procedure for finding best decisions (in design: pipe diameters, pump locations and sizes, etc.; in operation: which pumps should be on/off in each time period, valve settings, etc.) without going through a trial-and-error process. The optimization may use some results of network analysis as parameter values. The optimal solution may, in turn, need to be verified and possibly refined with the aid of network analysis (see Figure 4-1).

Before discussing formal optimization methods, it is worth emphasizing the very important role of network analysis and simulation in optimization studies. Steady-state network analysis and extended-period simulation (EPS), or operation-over-time, programs can be used to reduce the optimization model to manageable proportions and to double check the feasibility of model solutions. Reducing the problem and double checking the solution are necessary due to the complexity of the optimization

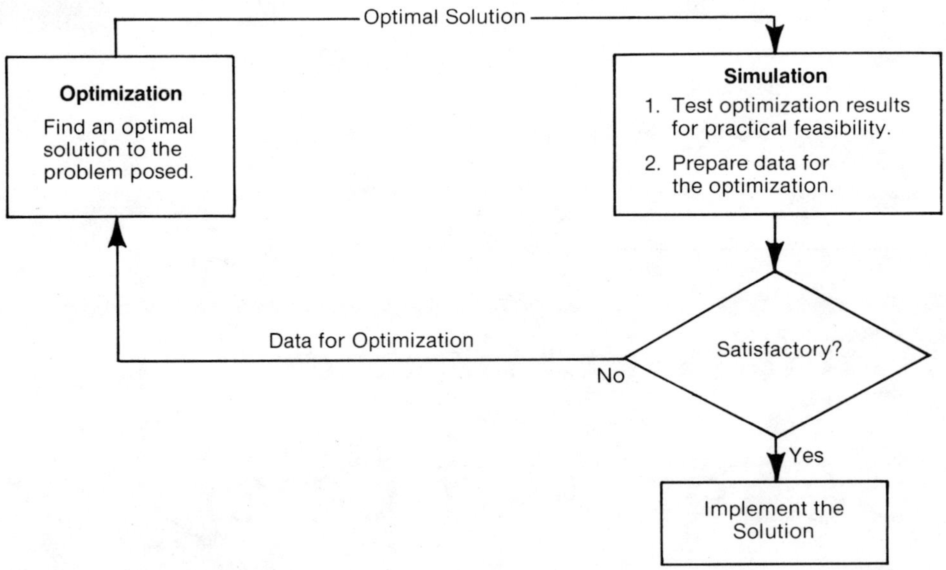

Figure 4-1 Flow-chart showing how optimization fits into the process of network analysis.

model. The complexity of the optimization model is due, in turn, to the complexity of distribution systems in general.

Design and operation of a distribution network involve

- complex physical systems of pressure zones, storage reservoirs, sources, booster pumps, control devices, and the pipeline network;
- uncertainty about future demands, such as long-term trends, variation over the day, week, and seasons, and fire flows; and
- complex requirements for quality of service, particularly pressures and reliability.

Therefore, there probably will not be, in the near or even in the more-remote future, computer programs for complete automatic optimization of design or operation. Instead, optimization programs are used to solve a well-defined reduced problem, one in which some decisions have already been made. Network analysis and simulation are used in the optimization process to help the analyst make these decisions by allowing him or her to

- gain familiarity with the existing network (if one exists) and with the expected conditions under which it will operate throughout its lifetime;
- evaluate alternatives for aspects of design and operation that will not be optimized explicitly by the optimization model; and
- examine the feasibility and practicality of optimization-model solutions and, if necessary, fine tune them.

Furthermore, some optimization methods use a network solver, or network-analysis software, as an integral part of the optimization program to guarantee the hydraulic feasibility of the optimization solution.

THE OPTIMIZATION MODEL

An optimization model is made up of four components, including decision variables, parameters, constraints, and the objective function. *Decision variables* are design and operational variables that are determined by solving the optimization model.

Examples of decision variables include pipe diameters, head added by a pump, and the time at which a pump will be turned on. *Parameters* are numerical coefficients and variables that have fixed values for each formulation of the optimization model. There is an exception: Parameters may be varied for sensitivity analysis. Examples of parameters include cost of materials, equipment and energy, and pipe-roughness coefficients. *Constraints* are conditions that the decision variables must satisfy. Some constraints are continuity of mass at a node and head-loss balance around a loop. The *objective function* is the criterion according to which various optimization-model solutions are ranked and the optimal solution selected. There may be more than one objective function for each formulation of the optimization model; in such cases, objective functions may compete. Examples of the objective function include total cost and energy consumed over a day.

An optimization model is established by finding the values of the decision variables ($x_1, \ldots, x_n$) that minimize (or maximize) the value of the given objective function $f(x_1, \ldots, x_n)$ while satisfying a given set of m constraints, expressed as equalities or inequalities. The following are some examples:

$$g_i(x_1, \ldots, x_n) = b_i$$

$$i = 1, \ldots, m$$

$$g_i(x_1, \ldots, x_n) \leq b_i$$

Frequently, there are several competing objectives, for example, minimum cost and maximum reliability. Methods exist for identifying good compromise solutions in which the trade-off between the objectives is the best in some sense. A description and discussion of such methods is beyond the scope of this manual; the interested reader is referred to Keeney and Raiffa[1] and Cohon.[2]

Objectives of Design and Operation

Two objectives should be achieved through the design and operation of a distribution system. The dominant objective is minimizing costs. Design affects life-cycle costs, including investment, operation, and maintenance over the life of the network. Real-time operation affects energy and operating costs. The second objective in design and operation, which is actually a whole set of objectives, is quality of service. This objective competes with the cost objective, that is, the higher the quality of the service, the higher the cost.

Quality-of-service objectives include reliability of supply, adequate pressures, and emergency supplies (that is, fire flows).

Reliability is, in itself, a composite and complex objective. Only recently has work begun on formalizing definitions of reliability and developing methods for assessing the reliability of supply systems.[3,4] Still missing, in all but relatively simple systems, are accepted measures of reliability and techniques for computing such measures. Considerations of reliability are usually introduced into optimization models as constraints, such as the requirement not to have dead ends and to close all loops in the system, or the specifications for stand-by pumps or for sizing storage. Much work must be done before reliability objectives and constraints can be incorporated into design and operation studies, including optimization, in a rational and systematic manner.

The trade-offs between cost and quality-of-service objectives other than reliability deserve more systematic study than has been conducted. Ideally, such studies should provide estimates of the value of quality of service, such as the potential economic loss due to less-than-perfect reliability, lower-than-specified pressures, and so on. Until such estimates become available, specification of quality-of-service criteria may be somewhat arbitrary.

Quality-of-service criteria should be considered constraints, and the analyst should strive to minimize the costs of these constraints. Thus, optimization models have cost minimization as the objective and quality-of-service criteria as constraints.

Constraints

Quality-of-service constraints include

- meeting all demands,
- maintaining pressures at all demand nodes above the specified standard (the specified standard may be different for average-day, peak-hour, fire-flow, and other demand conditions), and
- connecting each demand node by at least two pipes to provide reliability and to ensure that there are no dead ends.

Other constraints include physical laws and practical considerations. Physical laws include continuity of flows at nodes, continuity of the hydraulic grade line along any selected path in the network, maintenance of reservoir levels within limits, and head-discharge relationship in pipes and other hydraulic elements. Decision variables must be subject to practical considerations, such as commercially-available pipe diameters, the absence of negative pressures in pipes, and realistic reservoir sizes and configurations. Other types of constraints may result from the way in which the mathematical optimization model is formulated.

PLANNING, DESIGN, AND OPERATION

Planning a water supply and distribution system is the process of determining the basic structure and layout of the system, which sources will be used and in what order of priority, where water treatment plants and main storages should be located, what type of skeleton trunk system will be developed, and so on. Design is the process of selecting sizes for all facilities. Operation involves both operating policies, which determine how the controlled elements, such as pumps and valves, will be operated under design conditions (for example, for average-day and for peak-hour demands in the target year), and real-time operation, which determines the hour-by-hour operation of the system for one day to one week ahead.

Planning, design, and operating policies are integral parts of a single problem. When designing a system (in the general sense) the engineer determines the best plan, design, and operating policy. Real-time operation is treated as a separate problem.

OPTIMAL PLANNING, DESIGN, AND OPERATION

In an optimization model for a water distribution system, the decision variables include, among others, pipe diameter, material, and wall thickness; pump sizes; reservoir sizes; valve locations; and operation of pumps and valves for each of several typical (that is average daily or peak-hour) or emergency (that is, fire flow) demands.

Selecting the best layouts for a system is done by allowing some decision variables to take on zero values in the optimal solution. Thus, the optimization model can indicate a preferred plan by eliminating certain variables—if the model is allowed to do so and if it is not constrained to have a nonzero value for a certain variable. The model cannot, however, create an element where one was not present to begin with. Therefore, the optimization model can handle planning only in a very restricted sense. Most of the planning must be done by the engineer using creativity, experience, and judgment. The engineer remains the architect of the system; the optimization model is used to optimize the design of the engineer's plan.

Optimization for Branching Systems

Linear programming model. A branching system has no loops and is fed from a single source. When the demands are known, flows in all links of the branching system are also known. A linear programming (LP) optimization model, which has linear constraints and objective function, is used in planning a branching stem. The LP optimization model is developed by defining a new type of decision variable, the "length of segments of constant diameter."

To define this decision variable, select a set of candidate diameters for each pipe and denote by x_{ijm} the length of the segment of diameter m in the link between nodes i and j. These lengths must satisfy the following equation for all pipes:

$$\sum_{m} x_{ijm} = L_{ij} \qquad \text{(Eq 4-1)}$$

Where:

$\sum_{m}$ = summation over segments of given diameter m

L_{ij} = the length of the link.

The following equation is the hydraulic gradient in the segment:

$$J_{ijm} = \propto (Q_{ij}/CHW_{ijm})^{1.852} D_{ijm}^{-4.87} \qquad \text{(Eq 4-2)}$$

Where:

$\propto$ = a numerical coefficient that depends on the units used
Q_{ij} = flow through the link
CHW_{ijm} = the Hazen–Williams coefficient
D_{ijm} = the diameter of the segment.

Following a path from a node where the head H_s is known (for example, the source or a reservoir), the head at any selected node H_n can be computed by totaling the head losses between H_s and H_n. The constraint that requires this head to be within given bounds is

$$H\text{min}_n \leq H_s \pm \sum_{(i,j)} \sum_{m} J_{ijm} x_{ijm} \leq H\text{max}_n \qquad \text{(Eq 4-3)}$$

Where:

$H\text{min}_n$ = the minimum acceptable head at the node
$H\text{max}_n$ = the maximum acceptable head at the node
J_{ijm} = the hydraulic gradient in the segment (see Eq 4-2) (In figuring J_{ijm} for use in this equation, CHW_{ijm} is assumed to be known.)

$\sum_{(i,j)}$ = the summation over all links in the path that connects nodes s and n

$\sum_{m}$ = the summation over the segments of given diameter in the link.

The objective function is

$$\min F = \sum_{(i,j)} \sum_{m} C_{ijm} x_{ijm} \qquad \text{(Eq 4-4)}$$

Where:

C_{ijm} = the cost per unit length of pipe with diameter m in link (i, j), a cost that may vary over the network and, therefore, depends on the location of the link.

The summation is carried out over all segments of all links in the network.

Minimization, as illustrated in Eq 4-4, subject to the constraints in Eq 4-1 and Eq 4-3, is a linear program. It can be expanded to include pumps and to consider a number of loadings. The resulting model is

$$\min F = \sum_{(i,j)} \sum_{m} C_{ijm} x_{ijm} + \sum_{k} C(XPM_k) + \propto \sum_{k} \sum_{d} \beta(d) XP_k(d) Q_k(d) \qquad \text{(Eq 4-5)}$$

subject to the following constraints:

$$\sum_{m} x_{ijm} = L_{ij} \qquad \text{for all } (i, j) \qquad \text{(Eq 4-6)}$$

$$H\min_n(d) \leq H_s(d) \pm \sum_{k} XP_k(d) \pm \sum_{(i,j)} \sum_{m} J_{ijm}(d) x_{ijm} \pm \sum_{v} XV_v(d) \leq H\max_n(d) \qquad \text{(Eq 4-7)}$$

$$XPM_k \geq K \cdot XP_k(d) Q_k(d) \qquad \text{for all } d\text{, for each pump } k \qquad \text{(Eq 4-8)}$$

Where:

(i, j) = index of the link from node i to node j
m = index of the segment with diameter m
d = index of the loadings
k = index of the pumps
v = index of the valves
$C(XPM_k)$ = cost of a pump as a function of its size XPM_k
XPM_k = size of pump k, determined in Eq 4-8 as the maximum power required over all loadings
$\propto$ = coefficient for converting the product of head and flow into electric energy (assuming the overall pump efficiency) and then figuring the cash equivalent
$\beta(d)$ = present-value factor for loading d, used to convert the time stream of expenditures into present value, taking into consideration the planning horizon, the percentage of time in which loading d prevails, and the interest rate

$XP_k(d)$ = head added by pump k in loading d
$Q_k(d)$ = flow through pump k in loading d
$XV_v(d)$ = head loss in the valve for loading d
K = coefficient to convert the product of head and flow to pump size (for example, in horsepower).

NOTES:
The summations in the objective function (Eq 4-5) are over all pipes and pumps in the network.
The summation in Eq 4-7 is over only those elements that are on the path between nodes s and n.
The ± signs take into consideration the direction of flow.
All other variables have been previously defined.

The only nonlinear term in these equations is $C(XPM_k)$, and this can be handled in the LP as follows: Assume $C(XPM_k) = C \cdot XPM_k$, where C is a constant ratio of dollars to horsepower. Solve the LP, check the accuracy of this value of C at the resulting value of XPM, and, if necessary, adjust and re-solve the LP. This simple process converges well, because the unit cost is only mildly nonlinear against pump size.

Valves are added to the LP in Eq 4-7 in order to meet head constraints. Whenever possible, the LP optimization will eliminate valves because their presence increases cost and carries no benefit.

If this optimization model was formulated with pipe diameters as decision variables, the result would be a nonlinear model, because both head loss and cost are nonlinear functions of diameter. Using instead length of segments of constant diameter as decision variables allows the optimization model to be formulated as a linear program. This "trick" involves a few conditions that must be taken into consideration when formulating the model and solving the problem.

First, the list of candidate diameters should cover a range of diameters wide enough so that the limits of the list do not constrain the selection of diameters—except, of course, when the available diameters are, indeed, limited. If, in the solution of the LP, a link is made entirely of a diameter that is at one extreme of the candidate list, then the range of diameters should be expanded in that direction, and the LP should be augmented accordingly and re-solved.

Second, it can be proved that the optimal solution of the LP will have, at most, two segments for each link. Allowing a pipe to be made of two segments is reasonable and practical in many cases. If it is not—for example, if the solution indicates 480 ft of one diameter and 20 ft of another—then appropriate changes in the candidate list and/or constraint adjustments must be made, and the problem re-solved.

If the problem being solved is an expansion or modification of an existing system, then the pipes, pumps, and valves that are to remain unchanged are introduced into the model as fixed values and are not allowed to change.

To summarize, the steps for LP optimization of a new or expanding branching network are

1. Lay out the system and locate pumps and valves wherever they seem necessary or reasonable.
2. Determine flows in each link for each loading, or demand.
3. Select candidate diameters for each link. This may be done by assuming that the head loss gradient will be within a certain range (for example, 1–10/1000) for the range of flows in each pipe. This first proposed list will be expanded if necessary.

4. Using common sense and simple hydraulic reasoning, select the nodes at which minimum and/or maximum head constraints will be imposed (see Eq 4-7). The nodes are selected so that if the head constraints of the selected nodes are satisfied then, in all probability, the head constraints for all other nodes also will be satisfied. If this does not hold in the solution of the LP, then add a new head constraint wherever appropriate and re-solve the resulting LP.

5. Compute the coefficients $J_{ijm}(d)$, $\propto$, $\beta(d)$, and K. Obtain costs C_{ijm} (which, as the indices indicate, may vary with location) and estimate the unit-capital cost of pump horsepower.

6. Set up the LP and solve it. Iterate, as needed, according to considerations detailed above.

This optimization can be conducted by anyone who has a basic understanding of optimization and access to an LP code (which is readily available on both mainframes and microcomputers). Obviously, an understanding of hydraulics and of the system under study is also essential. Early work on branching systems was carried out by Karmeli et al.,[5] Gupta,[6] and Gupta et al.[7]

Dynamic programming model. The LP optimization problem solved above can also be solved by dynamic programming (DP). This procedure is somewhat inferior to LP in terms of computation, but it allows inclusion of certain additional decision variables—for example, pipe wall thickness, which determines the pipe's pressure-bearing capacity. For this reason, DP is briefly presented here. For more detail see Kally[8] or Shamir.[9]

To establish a DP optimization model, the network is divided into segments. A segment is defined between adjacent demand nodes so that the flow remains constant along the segment. If the segments are deemed too long or when it is necessary to consider specific conditions at certain points along the pipe—for example, low pressure at the top of a hill or high pressure at a low point—then additional nodes are defined.

Decision variables for DP optimization include pipe diameters and wall thickness, pump capacities, and pump and valve operation. Constraints require heads at nodes to be within given bounds. For simplicity of presentation, this example considers a single loading condition (to eliminate the index d), and the wall thickness and pipe roughness are fixed. The optimization is for pipe diameters, pump sizes, and pump operation.

The state variables (a term defined in DP models, see p. 77) are heads at nodes, and the nodes are the stages. Computation proceeds upstream, starting from the downstream end of each branch. The recursive equation figures the optimal cost of the part of the system downstream from a particular node, given the head at the node.

$$F^*_{j+1}(H_{j+1}) = \min_{D_k} \left[g(D_k) + f(H_{j+1}, H_j) + F^*_j(H_j) \right] \qquad \text{(Eq 4-9)}$$

Where:

$*$ = optimal value of the factor to which $*$ refers. In this case, the optimal cost.
j = the node at the downstream end of the branch
$j{+}1$ = the first node upstream of node j
k = the segment that connects j to $j{+}1$
H = head

$\min\limits_{D_k}$ = minimization over all admissible values of D_k

D_k = the diameter of segment k

$g(D_k)$ = the cost of segment k as a function of its diameter

$f(H_{j+1}, H_j)$ = the cost (or benefit, when it appears with a minus sign) associated with the link, as a function of the heads at its two ends. For example, the value of water delivered to a customer may depend on the pressure at the take-off point.

$F^*_j(H_j)$ = the optimal cost of the part of the system downstream of node j, given the head at the node. It is called the Bellman function (of stage j), or the "residual optimal remainder."

The minimization is over all admissible values of D_k, and is performed for each of a set of discretized values of the head H_{j+1} over its admissible range.

Whenever $H_j = H_j(H_{j+1}, D_k, Q_k)$, which is simply computed with a pipe head loss formula, is outside its admissible range, the examined D_k is rejected.

When a pump delivers from the upstream node $j + 1$ to its downstream neighbor j, the recursive equation becomes

$$F^*_{j+1}(H_{j+1}) = \min_{H_j} \left[g(H_{j+1}, H_j) + f(H_{j+1}, H_j) + F^*_j(H_j) \right] \quad \text{(Eq 4-10)}$$

Where:

$g(H_{j+1}, H_j)$ = the capital-plus-operating cost for a pump designed to deliver the known discharge Q_k from head H_{j+1} at its intake to head H_j at its discharge $(H_j > H_{j+1})$

Q_k = the known discharge of a pump k

At every branching node of the network, contributions from all downstream branches that join at the node are added together. At such a node j, $F^*_j(H_j)$ is still the optimal cost of the entire system downstream from that point, except that it is the sum over all branches that diverge from the node.

The computational scheme for optimization with a single loading is carried out as follows:

1. Start at downstream ends of the system. At each end node set $F^*_j(H_j) = 0$ for all values of H_j.

2. Move to the next node upstream. Name it $j + 1$. Discretize the head H_{j+1} over its permissible range using some reasonable increment. (An increment that is too small increases the number of necessary computations; an increment that is too large may cause a loss of accuracy.)

3. For each discretized value search over the range of available diameters D_k of the segment k that connects node $j + 1$ to its downstream neighbor, node j. For each D_k, compute H_j (the flow is known, and pipe roughness is selected in advance). If H_j is not within its allowed range, reject this D_k as infeasible, and continue the search. If H_j is within the allowed range, compute the sum in brackets in Eq 4-9: cost of the pipe segment, plus the cost or benefit resulting from the heads H_{j+1} and H_j (when no such cost or benefit exists, this term is set to zero), plus the "residual optimal remainder" $F^*_j(H_j)$ at the downstream node j for the resulting head H_j.

4. The search over all D_k results in finding the diameter that minimizes the sum. This minimum sum is stored as the equation $F^*_{j+1}(H_{j+1})$ = the best value, for the present value of H_{j+1}. Store also $D^*(H_{j+1})$ minus the diameter that gave the minimum.

5. For each discretized value of H_{j+1} repeat steps 3 and 4. Rename the node just completed j and move to the next node upstream, naming it $j+1$.

NOTE: Whenever two or more branches meet at node $j+1$, $F^*_{j+1}(H_{j+1})$ is the sum, over all the branches, of their individual contributions to the cost.

6. When a pump is being considered, its downstream (discharge) and upstream (suction) nodes are j and $j+1$, respectively. Equation 4-10 is used to find for each H_{j+1} the value of H_j that renders the sum in brackets a minimum. Store also the value of $H^*_j(H_{j+1})$ that resulted in the minimum.

7. At the source node s search over all of its discretized head values to find the head H_s for which $F^*_s(H_s)$ is a minimum. If H_s is given and fixed, no such search is needed. The optimal cost of the whole system is $F^*_s(H_s)$.

8. Follow the paths backward, from the source to the extremities, using the stored values of $F^*_j(H_j)$ and corresponding $D^*(H_j)$ or $H^*_{j-1}(H_j)$, to determine the optimal values of heads, diameters, and pumping heads everywhere in the system. An asterisk identifies an optimal value: F^* means optimal cost; D^* means optimal diameter; and H^* means optimal head.

The DP optimization method can be expanded to more than one loading condition, but it is practical for no more than two or three. Using more than one loading condition requires changes in the DP equations. Equation 4-9 still holds, if H_j and H_{j+1} are understood to be vectors of values—one for each loading at each node. D_k is still a single value. This means that the pairs of heads (H_{j+1}, H_j) for the various loadings must correspond to the same diameter D_k. In Eq 4-10, $g(H_{j+1}, H_j)$ must reflect the capital cost of a pump that can produce the head gain $(H_j - H_{j+1})$ for the given flow of each loading. (The head gain is different for each loading.)

There are no ready-made DP computer packages. To perform DP optimization, the analyst must formulate the optimization model and then write the computer program to solve it. While this is not too complex for someone with some experience in optimization, many water distribution engineers may require some expert help to implement such a procedure.

Optimization for Looped Systems

In contrast with branching systems, the flows throughout a looped system depend not only on the system demands and layout, but also on pipe diameters and on the operation of pumps and valves. To solve a simple optimization problem for a looped system, assume that the flow distribution throughout the network is known in advance for each loading (as it was in the branching system problem). Then the LP that was formulated for the branching system is appropriate for the looped system, provided a new set of constraints are added. For each loop, for each loading, add a constraint as follows:

$$\pm \sum_{(i,j)} \sum_{m} J_{ijm}(d)x_{ijm} \pm \sum_{k} XP_k(d) \pm \sum_{v} XV_v(d) = b_p(d) \qquad \text{(Eq 4-11)}$$

Where:

$\sum_{(i,j)}$ = the summation over all links in the loop

$\sum_{m}$ = the summation over the segments of given diameter in the link

J_{ijm} = the hydraulic gradient in the segment for loading d

x_{ijm} = the length of the segment of diameter m in the link between nodes i and j

$\sum_k$ = the summation over all pumps in the loop

$XP_k(d)$ = head added by pump k in loading d

$\sum_v$ = the summation over all valves in the loop

$XV_v(d)$ = head loss in the value for loading d

$b_p(d)$ = the head difference between the two ends of a path p, through the network. For a loop, which is a closed path, $b_p(d) = 0$. For open paths, as between two reservoirs with known heads, $b_p(d)$ is the head difference for loading (d).

Equation 4-11 must be written for all loops and all loadings in the system so that the resulting optimal solution is hydraulically balanced. This requirement expands the LP considerably and may lead one to consider other formulations of the optimization problem. Other formulations are discussed later in this chapter.

It is worth noting that even if the flow distributions for each loading are accepted as fixed—for example, taken from a conventional design of the system—it is possible to reduce the cost by formulating and solving the LP as outlined above.

The LP model with objective Eq 4-5 subject to Eq 4-6, Eq 4-7, Eq 4-8, and Eq 4-11 is for a given flow distribution throughout the network for each loading. After the LP has been solved, it is possible to extract from the solution information that can be used to modify the flow distribution so that an LP with the modified flow will yield a lower optimal cost.

Iterative implementation of the procedure for modifying the flows in the network, so that the cost is reduced successively, is somewhat involved and requires a working knowledge of linear programming. An outline of this method and the equations are presented here, but for a more complete guide to practical implementation refer to the original works (Alperovits and Shamir,[10] Shamir,[9] Quindry et al.[11]).

In the first iteration, the optimal value of the objective function F (Eq 4-5) is found for a specified flow distribution throughout the system, denoted Q. The objective function may be given simply as an initial guess, judged reasonable by engineering considerations, or as the result of a more detailed analysis. All that is required is that the flow distribution satisfy continuity at the nodes. The optimal value of F is found by linear programming and is for a particular flow distribution Q. This is represented by the following equation:

$$F = \text{LP}(Q) \quad \text{(Eq 4-12)}$$

Successive iterations use a set of changes in the flows, ΔQ such that when the LP problem is solved with the new flow distribution $Q + \Delta Q$, a lower value of F results.

$$\text{LP}(Q + \Delta Q) < \text{LP}(Q) \quad \text{(Eq 4-13)}$$

The number of flow changes possible equals the number of paths (Eq 4-11 must be satisfied for each path) times the number of loading conditions. The partial derivative of total system cost with respect to a change of flow in path p, denoted Q_p, is given by

$$\frac{\partial F}{\partial(\Delta Q_p)} = \pm \sum_r W_r \sum_{(i,j)\in p} \frac{1.852}{Q_{ij}} \sum_m \Delta H_{ijm} \qquad \text{(Eq 4-14)}$$

Where:

ΔQ_p = change of flow in path p

$\sum_r$ = the summation over all the head constraints

W_r = the dual variable, or "shadow price," of the constraint, and is obtained as part of the solution of the LP problem

$\sum_{(i,j)\in p}$ = the summation over all links of path p

Q_{ij} = flow in link (i, j)

$\sum_m$ = the summation over the segments m of the link

ΔH_{ijm} = head loss in segment m of link (i, j)

NOTE: The plus-or-minus sign has to do with the sense in which the constraint r is written relative to the positive direction taken for path p.

The partial derivatives of F with respect to all flow changes are combined into a gradient. The negative gradient points in the direction of steepest descent of F. A "step" is taken in this direction by changing the flow in the paths in proportion to their gradient components (derivatives). The step size can be controlled through specified parameters. More elaborate schemes can also be used for the search direction and step size (such as the BFGS method[12]), but details will be omitted here.

Another LP formulation for looped systems is based on assuming in advance the heads at all nodes. The design is optimized under this assumed condition and then the heads are adjusted to improve the solution.[11]

Using this formulation and given the heads at all nodes the following LP is formulated:

$$\min F = \sum_{(i,j)} B_{ij} L_{ij} x_{ij} \qquad \text{(Eq 4-15)}$$

subject to

$$\sum_i Q_{ij} = C_j \qquad \text{(Eq 4-16)}$$

for all nodes j.

Where:

B_{ij} = cost per unit length of pipe i, j per unit of x_{ij}

L_{ij} = length of pipe (i, j)

x_{ij} = $D_{ij}^{2.63}$, a converted variable, defined on the diameter

$\sum_i$ = summation over all nodes i connected by a pipe to node j

Q_{ij} = flow in pipe i, j, given by

$$Q_{ij} = \propto \bullet CHW_{ij} (\Delta H_{ij}/L_{ij})^{0.54} x_{ij} \qquad \text{(Eq 4-17)}$$

Where:

$\propto$ = a numerical coefficient that depends on the units used ($\propto = 6.2 \times 10^{-4}$ for flow in ft^3/s, diameter in inches, and ΔH and L in the same units)
CHW_{ij} = Hazen–Williams coefficient for the segment
ΔH = head loss in the pipe
L_{ij} = length of the segment

C_j = consumption at node j

After the LP is formulated, the cost function of pipes must be converted from dependence on the diameter D to dependence on a new variable $X = D^{2.63}$, and then linearized. A gradient of the total cost with respect to node heads is sought so that the heads can be changed to result in a lower cost when an LP using the changed heads is solved. The component of this gradient with respect to the head at node j is

$$\frac{\partial F}{\partial H_j} = 0.54 \left[W_j \left\{ \sum_{i \in I} \frac{Q_{ij}}{H_i - H_j} + \sum_{k \in K} \frac{Q_{jk}}{H_j - H_k} \right\} - \sum_{i \in I} W_i \frac{Q_{ij}}{H_i - H_j} - \sum_{k \in K} W_k \frac{Q_{jk}}{H_j - H_k} \right] \qquad \text{(Eq 4-18)}$$

Where:

W_j = the dual variable ("shadow price") of constraint for node j (see Eq 4-16)
I = the group of nodes from which flow goes to the node under consideration (in this case, j)
K = the group of nodes to which flow goes from node j

The gradient, given by its components in Eq 4-18, is used to change the heads.

Other optimization methods for looped systems combine an optimization routine with a network solver. Morgan and Goulter[13] use the basic method developed by Kally[14] to optimize a pipeline network fed by gravity with a given flow distribution. The new design is submitted to a network solver, which modifies the flows, and the resulting design is optimized. For a given flow distribution, the optimization model is

$$\min F = \sum_{j=1}^{NL} (C_{jdr} X_{jdr} + C_{jds} X_{jds}) \qquad \text{(Eq 4-19)}$$

subject to

$$\sum_{j \in P_i} (K_{ij} G_{jdr} X_{jdr} + K_{ij} G_{jds} X_{jds}) \leq H_i - h_i \qquad \text{(Eq 4-20)}$$

$$X_{jdr} \leq L_j \text{ and } X_{jds} \leq L_j \qquad \text{(Eq 4-21)}$$

Where:

NL = total number of pipes in the network
j = index of pipe
d = index of diameters
r = index of the diameter one larger than d
s = index of the diameter one smaller than d
C_{jdr} = cost of changing a unit length in j from diameter d to diameter r
X_{jdr} = decision variable: the length to be changed from diameter d to diameter r
C_{jds} = cost saving resulting from changing a unit length in j from diameter d to diameter s ($G_{jds} < 0$)
X_{jds} = decision variable: the length to be changed from diameter d to diameter s
P_i = path leading from the source to node i
K_{ij} = weight given to the link between nodes i and j (see Morgan and Goulter[13] for details)
G_{jdr} = change in hydraulic gradient caused by going from diameter d to diameter r with the given flow
G_{jds} = change in hydraulic gradient caused by going from diameter d to diameter s with the given flow
h_i = existing head at node i, before changes are made (i.e., in the last iteration of the network solution)
H_i = computed head at node i
L_j = length of pipe j

The constraints expressed in Eq 4-20 are formulated for each loading. Many of them can then be eliminated because they are redundant. Eliminating constraints reduces the size of the LP to be solved.[13] The method does not deal with pumping.

Other Optimization Methods

Several methods of optimization are in various stages of development or implementation.

Jeppson[15] proposes a set of guidelines for improving the design of a network using a network solver as the basic tool for analysis. Gessler[16] uses enumeration to examine many combinations of pipe sizes. Each combination is tested for its cost relative to the previous best solution. If it is an improvement, the network is solved and tested for hydraulic feasibility. If the pipe-size combination passes this test, it is saved as the best to date, and the enumeration continues. Lansey and Mays[17] use an optimization method that combines the general nonlinear optimization method GRG (generalized reduced gradient) with a network solver. Shamir[18] used this approach previously, but without a general-purpose GRG code. Ormsbee[19] uses a general-purpose optimization scheme—the complex method—combined with hydraulic analysis.

OPTIMAL REAL-TIME OPERATION

Real-time operation of a water distribution system comprises two complementary decision-making processes. The first is preparation of an operating plan at the beginning of each time period. The plan is based on forecasts of demands, equipment outages, and energy prices. The typical time period is one day. However, for systems with ample storage and slow changes the period might be extended to several days, while for systems with rapid, less predictable changes and/or little storage the period may have to be less than 24 hours. The other decision-making process involves modification and adjustment of the operating plan, if data collected over time indicate a significant change from the forecasts used in preparing the plan.

For distribution systems that operate much of the time under practically constant conditions, it may not be necessary to prepare a new operating plan every day. A fixed plan can be defined and remain valid as long as demands, energy tariffs, availability of the pumping equipment, and other similar factors do not vary much from the standard. A fixed plan can be hard wired into the system as a set of automatic controls on pumps and valves. This is called a control routine.

This section deals mainly with more advanced computer-controlled real-time systems, and with the software needed to compute the optimal control in particular. For distribution systems that do not have on-line computing power, the methods and software described here must be used off-line to prepare, in advance, control routines. These control routines are then given as instructions to the operators or implemented as hard-wired controls in the equipment.

Computer Control System

Reasons for use. A survey[20,21] by the AWWA Computer Assisted Design of Water Systems (CADWS) Committee indicates that many utilities would like to move towards computer-controlled or, at least, computer-assisted real-time operation. Many utilities routinely use computer programs for off-line studies of operation and of capacity expansion. The trend towards real-time control is motivated by some or all of the following factors, in combinations that vary among systems.

- Operation of the existing system is becoming increasingly more complex. New facilities are added and the existing ones are further interconnected. Sources of different qualities, reliabilities, and costs are introduced. Demands increase, and it is more difficult to meet them reliably. All of these changes require system-wide, rather than local, considerations in the preparation of operating plans.
- Retirement of experienced operators, who often are not replaced by people of similar capability or experience.
- Aging systems fail more frequently, leading to difficult operational decisions that must be made quickly under stress.
- High operating costs justify investments in attempts to optimize.
- Control and computer hardware are rapidly becoming cheaper, more available, and reliable.
- Computer control systems in treatment plants, power systems, and water systems are becoming more common, providing experience from which to learn.
- Managers, engineers, and operators are more comfortable and less threatened by computers.
- Mathematical models, optimization algorithms, and control software have developed considerably in recent years.

Components. A computer-based control system includes hardware, data base, and software, as follows:

- A means of measuring and transmitting field data to the control center.
- Pumps and valves that can be controlled remotely from the control center.

- One or more computers, operators' consoles, displays, printers, and other peripheral equipment.
- A data base for historical and current data, including network components, water consumptions, operational records, reservoir levels, operating status of pumps, and so on.
- Software for managing the data base. The software should perform standard statistical and mathematical analyses and display data and results in graphic and tabular form for the operators.
- Software for analyzing the data, for forecasting demands, and for determining good (optimal, if possible) operating plans. This is called the system software, or the control software.

There are two types of computer control systems. A system with all but the system software is a supervisory control and data acquisition (SCADA) system. A SCADA system has software to manage data collection, the data base, graphic and tabular screen displays of the system and data, and to transmit manually imposed controls from the control center to the proper locations in the field. Many SCADA systems exist, and more are being installed. However, the state-of-the-art control systems are distributed systems, which have a control center and remote units distributed in the field. The remote units have microprocessors; they can handle data and command traffic. They also provide local protection and control of the facility. Modern remote units accept down loading of their local control programs from the control center.

System software is necessary to computer control of a system. Only when such software exists can the system begin to move from a SCADA system to computer control. There are many levels of computer control, from occasional off-line analyses of system operation, which aid operators in evaluating their actions, to automatic control, where control loops are closed directly from the computer. Moving from SCADA to computer control should be done gradually, with caution. The ultimate degree of automation will vary among systems, depending on many considerations. But even when the final objective is a high degree of automation, the change is best made in stages. First, the analysis should be done off-line and the results presented to the operators to accept or reject. When sufficient experience and confidence in the computed operating plans is gained, one can begin to close loops: first, those that look safest to delegate to the computer, and then, with time, more and more. This method minimizes the possibility that the system will perform more poorly under computer control than it did under unassisted operator control. Thus, it is recommended to automate only proven algorithms and to operate with a person in the control loop as long as necessary, possibly always.

Control Objectives

Operating a system effectively means keeping operating costs low while meeting demands reliably, keeping reservoir levels and system pressures within bounds, and possibly satisfying some practical operating considerations as well. Various components of the system software contribute to effective operation; these components and their contributions are discussed in the following sections.

Demand forecasting. The first condition for effective operation is the ability to estimate future demands for a period of one or more days. Operators do this intuitively, from experience. They use such information as day of the week, hour of the day, weather, special events (such as a holiday, or a major sporting event), last day and last hour consumptions, and so on. Some cities use rain gauges to monitor rainfall throughout the system area. Operators may find this information necessary, particularly where lawn irrigation is an important component of consumption.

A demand-forecasting program uses past demand data and may use additional information, such as weather forecasts, to forecast demands 24 h ahead. When necessary, a weekly forecast is also made. A variety of techniques that can be used to forecast demand have been developed. Based on methods previously developed—a Kalman filter by Perry[22] and a Fourier transform plus exponential smoothing by Gray[23]—Moss[24] developed a six-harmonic Fourier predictor with triple exponential smoothing. This program performs well for the urban demands in the East Worcesterishire Water Company in England. Prior to Moss's work, Shimron and Shamir[25] tried two simpler time-series analysis methods to forecast demands in Israel. Recently, Maidment and co-workers[26] developed a set of procedures for forecasting daily demands with response to rainfall.

It seems that a combination of time-series methods and whatever relevant information can be obtained directly from consumers and about weather forecasts is probably the most promising.

Network analysis and simulation. Knowing how different controls will cause the hydraulic system to react, in terms of its physical performance and the resulting cost of operation, is essential to effective operation. As with demand forecasting, operators draw on their experience and understanding of the system. A good operator can predict, to a reasonable approximation, the system's reaction to any control and can estimate how expensive that control is compared to others. However, such predictions may be less accurate or even seriously in error as the hydraulic system becomes more complex, especially when the system is in a state unfamiliar to the operator (for example, due to some major failure).

Hydraulic simulation. An essential component of the system software is a hydraulic simulator, a program that computes the behavior of the hydraulic system over time under a given control routine or operating plan. The simulator is used for a variety of purposes including

- to study the system, validate the model specified and data used, understand the hydraulic response to various loadings and operating options, and compute operating costs for proposed control routines and operating plans.
- to help in developing a more aggregate model of the hydraulic system, if needed for the optimization model. This involves elimination of details, combination of elements into equivalents, and computation of functional forms and coefficient values for the aggregate model.
- to test the results of optimal-operation models for hydraulic feasibility and for accuracy of the cost calculations.

Network analysis and simulation can be very effective tools to aid operators in making decisions. Operators submit proposed operating plans, and the programs simulate the physical results and costs. Operators can then try different controls and select the best plan among those tried. This can be done in training sessions (off-line, so to speak), as well as in real-time operations.

The Optimal-Control Problem

The optimal-control problem for a water-supply and distribution system is how to operate the pumps and valves so as to minimize the total cost while meeting demands and satisfying minimum pressure constraints. The rest of this chapter discusses formulation of a sample optimal-control problem and various methods for solving it.

Demands. It is assumed in this formulation that demands are given and must be met. A broader formulation might include demands as decision variables. The objective would then be to maximize net benefit, that is, the difference between the benefits resulting from supply and the costs. This broader formulation might be relevant in irrigation systems and possibly for industrial consumers, but it is outside the scope of this discussion. The optimal-control problem discussed in this chapter deals with urban supply, and demands are irrelevant to the decision problem.

Time span. For most optimal-control problems, the time horizon, or span of time, under consideration is 24 h, because demands repeat approximately on a daily cycle. Occasionally, the weekly behavior must be considered in some way, as well. This occurs when there is weekly storage, that is, reservoirs that fill over the weekend and are slowly depleted over the week. This discussion concentrates on the daily operation problem, because weekly storage can be dealt with by a relatively simple modification.

The optimal-control problem in this chapter deals with hourly operation over one day. The approaches and methods that will be used are equally applicable to shorter or longer time periods—from a fraction of a day up to one week. The 1-h intervals may also vary. Fifteen- or 30-min intervals may have to be used during times of rapid change and longer time periods—2, 4, 6, or more hours—may be quite adequate under certain circumstances.

For a number of reasons, the control problem must be considered for the entire 24-h period and cannot be separated into individual hourly decisions. The primary reason is storage, which changes over time and links adjacent time intervals. Other reasons for linking the hourly problems over time are constraints on the minimum time elapsed between turning pumps on, off, and on again to prevent excessive wear and the fact that total operating time or energy consumption by a pump may influence costs.

The optimal-operation problem is formulated as an open-loop control for the 24-h period, with demands at each node given for all hours. The initial state is the reservoir volumes at present, and the final state is assumed known. Frequently, the problem is formulated for a period starting and ending at 6:00 a.m., with all reservoirs assumed to be full at both initial and final states. If a reservoir provides weekly storage, then its final volume at the end of the day can be lower than the initial volume by the amount corresponding to one day's withdrawal.

The control problem need not be solved for an entire 24-h period. Given the initial state at any time, the demand forecast for the remainder of the day and the final state, the control problem can be formulated. Because of the computational difficulties, which will be discussed later, present-control algorithms usually are solved once a day (usually at night). Some present-control algorithms have a procedure for modifying the operating plan during the day for the remaining time if actual conditions deviate from the forecast. This correction is computed by approximate methods, which makes them feasible for on-line computations.

Constraints. Solution of the optimal-control problem must satisfy constraints of several types. Demands must be met, source capacities must not exceed their upper limit, reservoir levels and pressures at nodes must be within their admissible ranges, and hydraulic laws of the system must hold. Hydraulic laws relevant to the optimal-control problem include continuity at nodes; head-discharge relations in pipes, pumps, and valves; and level–storage relations in reservoirs.

The resulting control problem is difficult to solve because the problem may be rather large and some controls are discrete (for example, pumps may be on or off). In addition, the network equations are nonlinear, and their solution for each time interval must be by iterative methods. There is, therefore, no way to obtain explicit expressions for dependent variables (such as flows into and out of reservoirs or pressures at nodes) as a function of the controls. Several approaches exist for dealing with these difficulties. Decomposition, aggregation, and iterative dynamic-programming methods can deal with the size of the problem. When pumps can be operated in many combinations to cover the range of possible discharges, the corresponding control can be considered continuous rather than discrete. Finally, network equations can be made linear and pressure constraints can be ignored (it is expected that, despite these changes, the optimal controls will result in feasible pressures).

Solution Approaches and Algorithms

The control problem can be solved by DP when the problem's dimensions do not make it computationally impractical. Dynamic programming can handle all of the aspects of the control problem: discrete control, constraints on the state variables (reservoir levels), and pressures. However, DP works well only when the number of state variables is small (ideally one or two reservoirs, but in any case no more than four).

The Hitachi Company in Japan has developed a method for determining operating policies so that the resulting pressures throughout the system closely match a prescribed pressure map. This does not fall into the same category of optimization schemes as discussed above, and there is not yet sufficient information about the method and its implementation to discuss its use.

$t = 1,\ldots,T$ = time periods, or stages for the DP (for operation over 1 day, $T = 24$)
$D(t)$ = vector of demands at nodes
$V(t)$ = vector of reservoir volumes at beginning of time t (These are the state variables of the DP.)
$P(t)$ = vector of pressures at nodes
$U(t)$ = vector of controls for pumps and valves
$f_t(V,U)$ = cost of operating the system during time period t, starting at reservoir volumes V, and applying controls U.

The optimization problem is

$$\min_{U(t)} \sum_{t=1}^{T} f_t\,[V(t), U(t)] \qquad \text{(Eq 4-22)}$$

The optimization problem is subject to constraints on reservoir volumes (or the corresponding levels), as shown in the following equation:

$$V(t) \le V(t) \le V(t) \qquad \text{(Eq 4-23)}$$

where volume changes are given by the following network dynamics:

$$V(t+1) = g\,[V(t), D(t), U(t)] \qquad \text{(Eq 4-24)}$$

Eq 4-24 is the result of solving the network equations with the reservoirs at volumes $V(t)$, the demands given as $D(t)$, and the controls $U(t)$ applied. This gives the new volumes in the reservoirs at the end of the time period $V(t+1)$.

In addition to constraints on reservoir volumes, there may also be constraints on pressures at some or all nodes. These are represented by the following equation:

$$P(T) \leq P(T) \leq \overline{P}(T) \qquad \text{(Eq 4-25)}$$

where pressures are given by solving the network equations

$$P(t) = h[V(t), D(t), U(t)] \qquad \text{(Eq 4-26)}$$

The beginning and end states $V(1)$ and $V(T+1)$ are given and $U(t)$ belongs to an admissible set of controls.

The recursive equation of the DP algorithm is

$$F^*_t[V(t)] = \min_{U(t)} \left\{ f_t[V(t), U(t)] + F^*_{t+1}[V(t+1)] \right\} \qquad \text{(Eq 4-27)}$$

Where:

$F^*_t[V(t)]$ = the Bellman function, the optimal cost of operating the system from the beginning of time period t, for the rest of the time horizon, starting with state $V(t)$.

The recursive equation is applied backward, starting from the final state $V(T+1)$, which is assumed known (for example, equal to the initial state so that the operation is cyclic), with $f^*_{T+1}[V(T+1)] = 0$. When $t = 1$ is reached, Eq 4-27 is applied only for the known value of $V(1)$. The optimal state trajectory—that is, the optimal sequence of reservoir volumes for each hour over the day—is recovered by computing the progress from the volumes at (t) with the corresponding optimal controls, for $t = 1, 2, \ldots, T$.

When applying the recursive equation backwards, for each $V(t)$ and $U(t)$ examined the network equations are solved, and the resulting $V(t+1)$ and $P(t)$ are computed. If $V(t+1)$ and $P(T)$ do not satisfy their constraints, as expressed in Eq 4-23 and Eq 4-25, respectively, then the control $U(t)$ is rejected as infeasible.

The volumes in the reservoirs are discretized over their permissible range. The volume increment for the discretization must balance between accuracy (many small increments), and computational efficiency (few large increments). Consider, for example, three reservoirs, each with a total depth of 5 m. Taking increments every 10 cm gives 50 increments per reservoir, or 50^3 = 125,000 discrete points for each of the 24 state vectors $V(t)$. This is the number of times Eq 4-27 is performed, and each time all possible controls (combinations of pumps and valves operated) must be examined. Furthermore, once the optimal control $U^*[V(t)]$ has been found for the state value $V(t)$ under consideration, then this control, as well as the value of the Bellman function $F^*_t[V(t)]$, must be stored. Thus, a fine discretization increases both computation time and computer memory requirements.

The number of reservoirs, which is the dimension of the state vector, is the primary cause for these computational difficulties. This is the "Curse of Dimensionality" of dynamic programming. Methods of aggregation (of several reservoirs into an equivalent one), successive approximations, and decomposition–coordination attempt to deal with this difficulty, as will be explained later.

Since the entire Bellman function is stored during computation for all possible values of $V(t)$, then as long as the demands do not change there is no need to recompute when $V(t)$ is found to deviate from the computed optimal volumes at time t. If the forecast $D(t)$ deviates markedly from that used initially, the optimization for the remaining time may have to be rerun. Dreizin et al.[27] used this procedure for a relatively simple system. For a larger system, the procedure proved computationally infeasible, and they resorted to a search procedure.

Variations on the DP formulation. Sterling and Coulbeck[28,29] tried to push the DP method to the limit by simplifying some of the equations, but the results were not very realistic.

Cohen and co-workers[30,31,32] developed two devices to get around the "Curse of Dimensionality": decomposition–coordination and aggregation. Space does not permit a full discussion of the details, so only the principles will be described.

Decomposition–coordination. In decomposition–coordination, the entire system is divided into non-overlapping subsystems. Each subsystem must be small enough to be solved by DP (as is or after some aggregation of its state vector, as explained below). Each subsystem has one or, at the most, two reservoirs. Assume the "cuts" between the subsystems are in pipes. Then, if the 24-h trajectory of either head or flow at the cut and the trajectory of cost of the other variable at the same point are given, the subnetwork's optimal-control problem can be solved. The solution specifies the internal controls of the subsystem and the trajectories of the variables that were free to change at the cut.

For example, consider adjacent subsystems A and B, and the following assignment of values at the cut between them: A "gives" B the trajectories (24 hourly values) of heads and costs of the flows, while B "gives" A trajectories of flows and costs of heads. First, fix the trajectories of flows and head costs at the cut; then, optimize A. The results include a trajectory of heads at the cut. Moreover, the cost of meeting each of the specified flows at the cut can be computed. To compute the cost, use the stored Bellman function for A and a simplified estimate of the change in the internal state vector (reservoir storage) due to an incremental change in one value of the specified flow trajectory at the cut. The resulting head and flow-cost trajectories are passed over to B, which can be optimized internally.

The results of each subsystem's optimization are thus trajectories of one physical variable (for example, head) at each of its "cut" connections to neighboring subsystems and of the cost of the other variable (in this case, flow) for each hour at the same connecting points. This information is passed on to the neighboring subsystems, which are similarly optimized. Iterations through the system proceed until convergence is achieved.

The coordination process works best when controls are continuous, and when the optimal control of a subsystem is not too sensitive to the boundary conditions at the cuts.

Location of the cuts and selection of the variable to be computed are something of an art, based on mathematical considerations and an understanding of system hydraulics. These decisions are far from trivial and require experience, judgment, and acquaintance with the network under study.

It may happen that some subsystems have two or even three reservoirs. A way to reduce the dimensionality of the subsystems' DP is to aggregate reservoirs. This means lumping two or more into one equivalent reservoir. This can be done without too much loss of accuracy only if the individual reservoirs operate naturally in tandem. This is the case when they are close and strongly connected, but can also occur when the closeness in behavior is less apparent until a hydraulic analysis of the network reveals it.

This procedure[32] has been implemented in a regional system west of Paris, France. In addition to the computational scheme outlined above, which is carried out off-line every night, it includes a procedure to correct the operating plan during the day if deviations from forecasts are noticed. This is done on-line, using the stored Bellman function from the off-line optimization.

Pressure-zones model. Zessler and Shamir[33] developed another iterative method, using DP as the basic problem-solving technique. Their method uses an iterative DP procedure to optimize the operation of systems that can be modeled as a number of pressure zones connected at reservoirs. The whole process is iterative at two levels: over time and, for each time step, over subsystems. Each level is carried to convergence, which is guaranteed mathematically only if the cost functions are convex.

In this procedure, the system is divided into subsystems with some overlap between them. The time-varying problem is solved by the approximate DP method known as progressive optimality. Progressing one time step (hour) at a time, the starting states $V(t-1)$ and end states $V(t+1)$, two time steps apart, are assumed fixed. Only $V(t)$ is allowed to vary, and its optimal value is located by a simple search through the admissible controls. After the optimal value of $V(t)$ has been determined by the optimization, it is fixed, together with $V(t+2)$, and $V(t+1)$ is optimized. This proceeds along the time axis from $t = 2$ to $t = T-1$, cyclically.

For each pair of time steps, the algorithm goes over the subsystems in sequence, optimizing each separately, with the others assumed fixed. This is done iteratively.

In this procedure, the pumping stations are represented by a function that gives the cost of pumping as a function of discharge. This function is defined with the aid of a network solver and then run for a range of operating conditions at each reservoir.

This algorithm was originally implemented on a very small microcomputer for one regional water supply system with seven subsystems, and it was tested extensively off-line. Subsequently, the program was generalized, written on a personal computer, and tested for other regional systems.

Other methods. Fallside and others[22,24] have developed and installed[34] a control system with a demand forecasting program for the East Worcesterishire Water Company (EWWC) in England. This system has been operating since the early 1970s and has had several major updates and improvements.

The procedure was developed specifically for EWWC and makes some assumptions about interzonal water transfers. The system is divided into six zones. In each zone there is one large reservoir, and the entire demand of the zone is lumped into one time-varying quantity. System dynamics for zone i are represented by the following continuity equation:

$$V_i(t+1) = V_i(t) + U_i(t) - D_i(t) - \sum_j Q_{ij}(t) \qquad \text{(Eq 4-28)}$$

Where:

V_i = volume in reservoir i
U_i = pumping production in the zone
D_i = demand of the zone
Q_{ij} = transfer from zone i to zone j.

It is assumed that Q_{ij} is fixed in advance; therefore, Q_{ij} is not optimized. This is suitable for the EWWC system but may not be elsewhere, where interzonal transfers are important decisions in the optimization. Having made this assumption, a special-purpose optimization process called "pump-priority logic" was evolved to determine the optimal controls.

Coulbeck[35] linearized the system equations as follows:

$$V(t+1) = A_1 V(t) + B_1 U(t) + C_1 D(t) \quad \text{(Eq 4-29)}$$

$$P(t) = A_2 V(t) + B_2 U(t) + C_2 D(t) \quad \text{(Eq 4-30)}$$

Where:

$A_1, B_1, \ldots, C_2$ are matrices of coefficients, which are obtained through extensive network simulation

V = reservoir volumes
U = the vector of controls
D = demand
P = pressures at nodes

Coulbeck considered in the cost function, the variable cost of energy over time, and a maximum energy demand tariff. The last component complicates the optimization process.

The Hitachi Company in Japan has developed a method for determining operating policies so that the resulting pressures throughout the system closely match a prescribed pressure map. This does not fall into the same category of optimization schemes as those discussed above, and there is not yet sufficient information about the method and its implementation to discuss its use.

References

These references do not cover all past work on optimization. The papers by Shamir[9] and by Walski[36] review the field and contain more extensive reference lists.

1. KEENEY, R.L. & RAIFFA, M. *Decisions With Multiple Objectives: Preferences and Value Tradeoffs*. John Wiley & Sons, New York (1976).
2. COHON, J. *Multiobjective Programming and Planning*. Academic Press, New York (1978).
3. SHAMIR, U. & HOWARD, C.D.D. Water Supply Reliability Theory. *Jour. AWWA*, 73:7:379 (1981).
4. — — —. Reliability and Risk Assessment for Water Supply Systems. In *Computer Applications in Water Resources* (H. Torno, ed.), 1218–1228. ASCE, New York (1985).
5. KARMELI D.; GADISH, Y.; & MEYERS, S. Design of Optimal Distribution Networks. *Jour. Pipeline Div.—ASCE*, 94:PL1:1 (1968).
6. GUPTA, I. Linear Programming Analysis of a Water Supply System. *AIIE Transactions*, 1:1:56–61 (1969).
7. GUPTA, I.; HASSAN, M.Z.; & COOK, J. Linear Programming Analysis of Water Supply Systems with Multiple Supply Points. *AIIE Transactions*, 4:3:200–204 (1972).
8. KALLY, E. Pipeline Planning by Dynamic Computer Programming. *Jour. AWWA*, 61:3:114–118 (1969).
9. SHAMIR, U. Optimization in Water Distribution Systems Engineering. Mathematical Programming Study, No. 11, 65–84. North Holland, New York (1979).
10. ALPEROVITS, E. & SHAMIR, U. Design of Optimal Water Distribution Systems. *Water Resources Research*, 13:6:885–900 (1977).
11. QUINDRY, G.E.; BRILL, E.D.; & LIEBMAN, J.C. Optimization of Looped Water Distribution Systems. *Jour. Envir. Engrg. Div.—ASCE*, 107:EE4:665–679 (1981).

12. GILL, P.E.; MURRAY, W.; & WRIGHT, M.H. *Practical Optimization*. Academic Press, New York (1981).
13. MORGAN, D.R. & GOULTER, I.C. Optimal Urban Water Distribution Design. *Water Resources Research*, 21:5:642–652 (1985).
14. KALLY, E. Computerized Planning of the Least Cost Water Distribution Network. *Water and Sewage Works*, R121–R127 (1972).
15. JEPPSON, R.W. Practical Optimization of Looped Water Systems. In *Computer Applications in Water Resources* (H. Torno, ed.), 723–731. ASCE, New York (1985).
16. GESSLER, J. Pipe Network Optimization by Enumeration. In *Computer Applications in Water Resources* (H. Torno, ed.), 572–581. ASCE, New York (1985).
17. LANSEY, K. & MAYS, L.W. A Methodology for Optimal Network Design. In *Computer Applications in Water Resources* (H. Torno, ed.), 732–738. ASCE, New York (1985).
18. SHAMIR, U. Optimal Design and Operation of Water Distribution Systems. *Water Resources Research*, 10:1:27–36 (1974).
19. ORMSBEE, L. OPNET: A Nonlinear Design Algorithm for Hydraulic Networks. In *Computer Applications in Water Resources* (H. Torno, ed.), 739–748. ASCE, New York (1985).
20. AWWA Computer Assisted Design of Water Systems Committee. Network Analysis Survey—1984. Proc. AWWA Ann. Conf., Dallas, Texas (June 1984).
21. VELON, J.P.; CESARIO, A.L.; & SHAMIR, U. Network Analysis of Treated Water System. AWWA Distribution Systems Symposium Proc., Syracuse, N.Y. (September 1984).
22. PERRY, F. An On-Line Optimal Control Scheme for Water Supply Networks. Doctoral dissertation. Cambridge University, Cambridge (1975).
23. GRAY, D.F. Use of Consumption Predictors. Cambridge Univ. Engrg. Dept. Tech. Rept. 4/78. Cambridge University, Cambridge (1978).
24. MOSS, S.M. On-Line Optimal Control of a Water Supply Network. Doctoral dissertation. Cambridge University, Cambridge (1979).
25. SHIMRON, Z. & SHAMIR, U. Short Range Forecasting of Water Demands. *Israel Jour. of Tech.*, 11:6:423–430 (1973).
26. MAIDMENT, D.; MIAOU, S.P.; & CRAWFORD, M. Response of Daily Water Use to Rainfall in Austin Texas. Dept. of Civil Engrg., Univ. of Texas at Austin (1984).
27. DREIZIN, Y. ET AL. Evaluation of Energy Savings in Regional Water Systems. (in Hebrew), Rept. 131.105. Mekorot Water Co. (April 1971).
28. STERLING, M.J.H. & COULBECK, B. Optimization of Water Pumping Costs by Hierarchical Methods. Proc. Institution of Civil Engrs., Vol. 59, 789–797 (1975-a).
29. — — —. Dynamic Programming Solution to Optimization of Pumping Costs. Proc. Institution of Civil Engrs., Vol. 59, 813–818 (1975-b).
30. JOALLAND, G. & COHEN, G. Optimal Control of a Water Distribution Network by Two Multilevel Methods. *Automatica*, 16:83–88 (1980).
31. COHEN, G. Optimal Control of Water Supply Networks. In *Optimization and Control of Dynamic Operational Research Models*, (S.G. Tzafetas, ed.), 251–276. North Holland, New York (1982).
32. CARPENTIER, P. & COHEN, G. Decomposition, Coordination and Aggregation in the Optimal Control of a Large Water Supply Network. Proc. IFAC World Congress. Budapest, Hungary (July 1984).
33. ZESSLER, U. & SHAMIR, U. Optimal Operation of Water Supply Systems. Unpublished.
34. MARLOW, K. & FALLSIDE, F. A Planned Strategy for Telemetry at the Water Works. In *Control and Instrumentation*, 47–51 (Dec. 1980).
35. COULBECK, B. Optimal Operations in Non-linear Water Networks. *Optimal Control Methods and Applications*, 1:2:131–141 (1980).
36. WALSKI, T.M. State-of-the-Art Pipe Network Optimization. In *Computer Applications in Water Resources* (H. Torno, ed.), 559–568. ASCE, New York (1985).

Chapter 5

Graphics Support Software

Computer graphics play an important role in supporting the decision-making process for planning and analyzing water supply and distribution systems. Coupling graphics with the output of network-analysis programs creates a visual representation that can lead to better assessment of system performance under various design and operating conditions.

ENHANCING NETWORK MAPPING

There are several ways computer graphics can be used to combine the presentation or documentation of maps with the results (flow, velocity, pressure, hydraulic-grade-line values, pipe-friction coefficients, and so on) of network-analysis programs. One approach is to integrate system mapping functions with the analytical program data base. In this way, the system is documented in detail, showing pipes, valves, tanks, pump stations, hydrants, service connections, meter locations, and other similar items. The pipe network can be shown as a separate layer or as part of underground maps. The maps can be supplemented further with surveying data and can be tied to a coordinate system, such as the state plane coordinate system. This graphic presentation of data parallels practices used by most water departments and is similar to those used for sectional plat records.

Management of graphics data as related to water lines usually falls under the jurisdiction of water departments or engineering departments. The pipe files can be used by departments in charge of the mapping system and can be presented as part of an overall underground records data base. The results obtained from the network-analysis program can be tied to the pipe files to show flow and velocity in each pipe and pressures (or HGL values) at each of the nodes under some standard level of demand and estimated pipe-friction coefficients.

A distribution network may be presented as part of the underground record map, which also includes right-of-ways, property lines, and streets (Figure 5-1). In addition, zoning patterns, the sewer system and other utilities, valve records, pump-

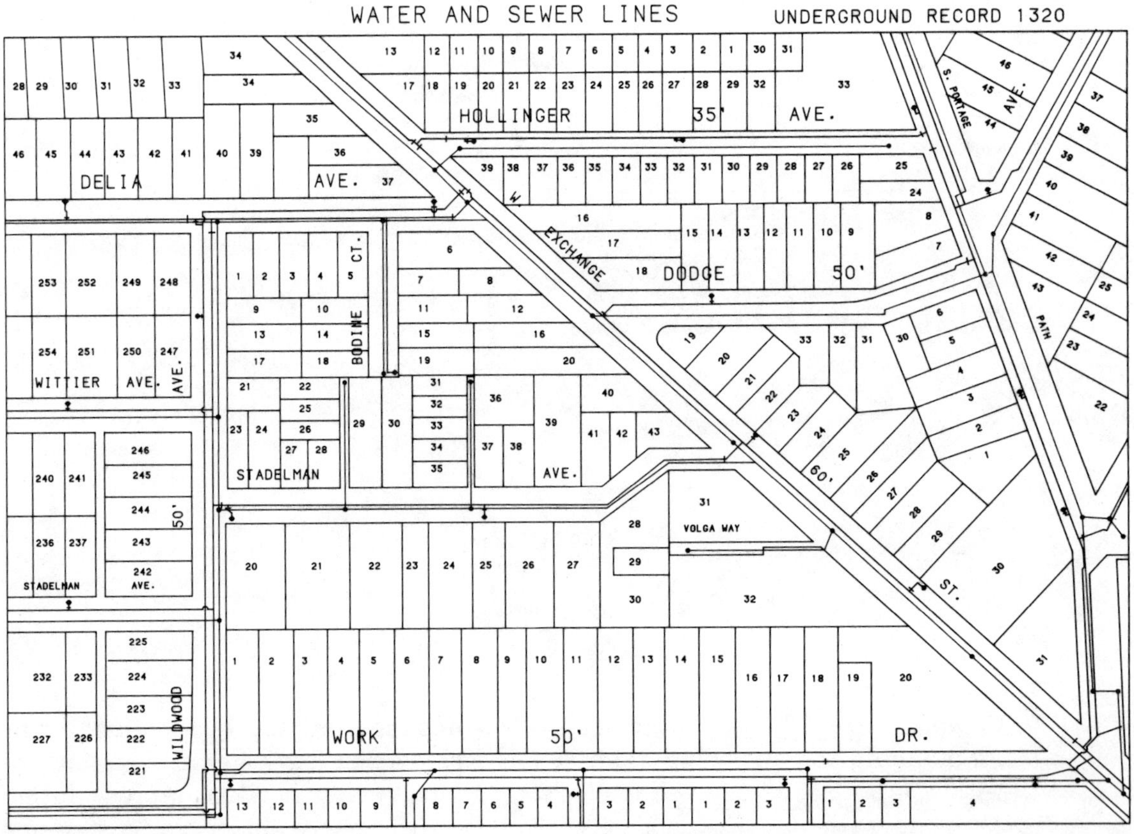

Figure 5-1 Underground record map.

station details and their operational characteristics, tank configurations, tank filling and emptying schedules, and hydrant-flow capability may be shown as part of the underground record map system data base. This information can be stored in layers, as shown schematically in Figure 5-2. The user may then retrieve only data of particular interest, for example, the layer containing water lines, valves, tees, hydrants, and street right-of-ways (Figure 5-3). More detailed information can be found on the valve-record layer (Figure 5-4). This layer shows the detail of a street intersection, with each hydrant and valve identified and located in relation to property boundaries. The normal operating state of each valve (open or shut) and the capacity of hydrants as determined from the hydraulic-analysis program may also be included. Such centralized record storage provides for effective management control, availability of consistent information to all users, and greater efficiency in the maintenance of records.

Another way to display results of hydraulic-analysis programs is to show each pipe as a line segment with street junctions or hydrants as end points (Figure 5-5). Some of the input data used by analysis programs, such as pipe diameters, pipe identification numbers, and node identification numbers, may be documented on these maps. Output data, such as pressures, HGL values, flows, and velocities, may be printed for each node or pipe. Pressure contours and HGL contours may be generated and superimposed on the pipe-network plot. In this way, the general pressure patterns developed throughout the system under a given set of supply and demand

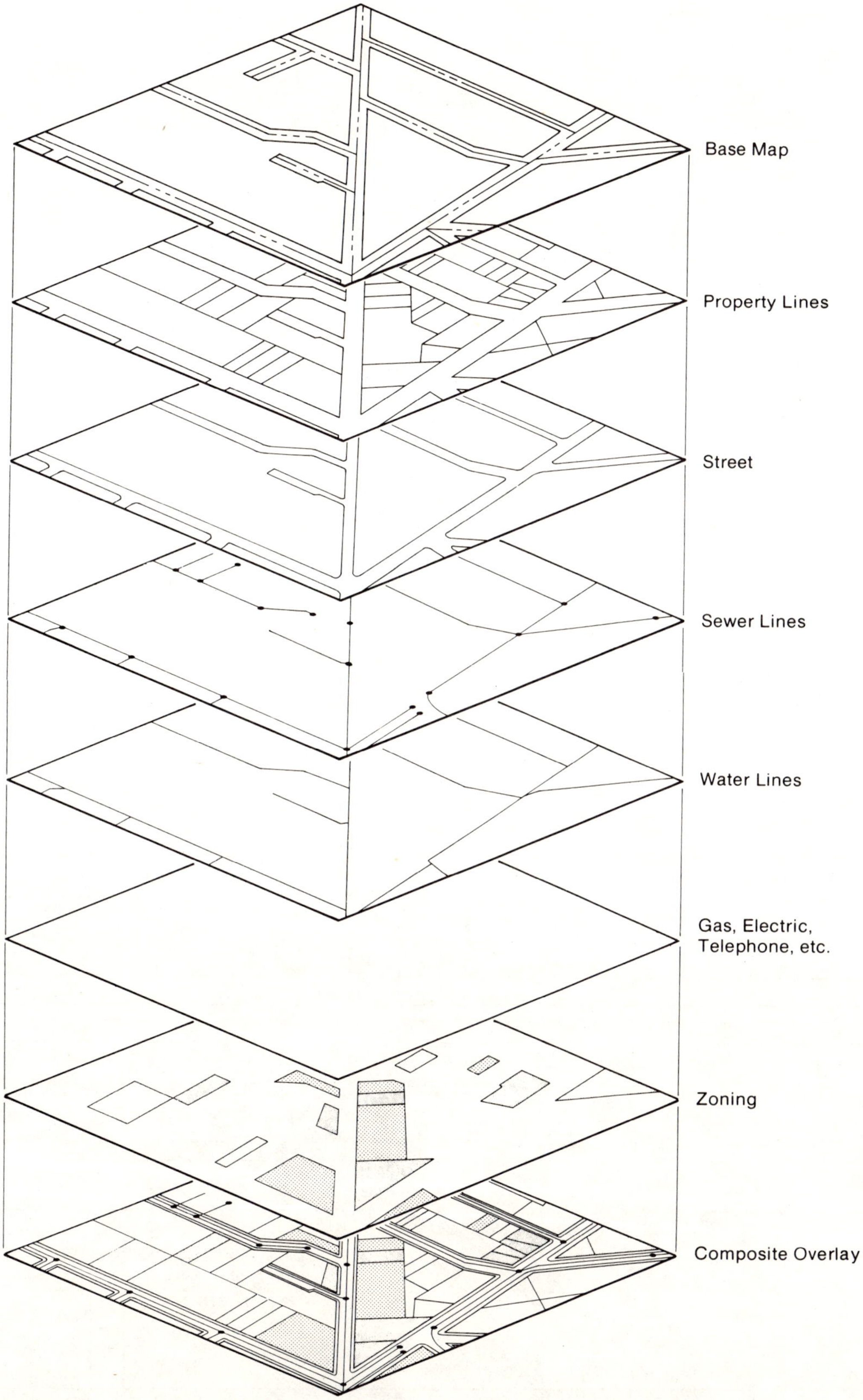

Figure 5-2 Layered mapping system.

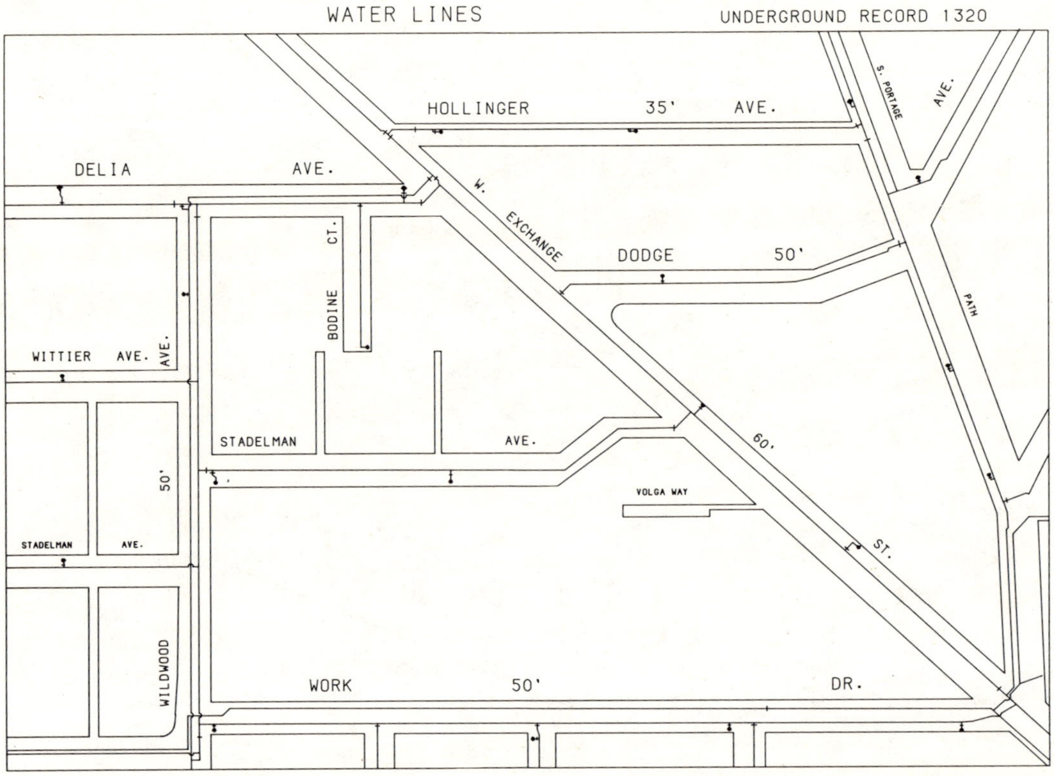

Figure 5-3 Map layer showing right-of-ways.

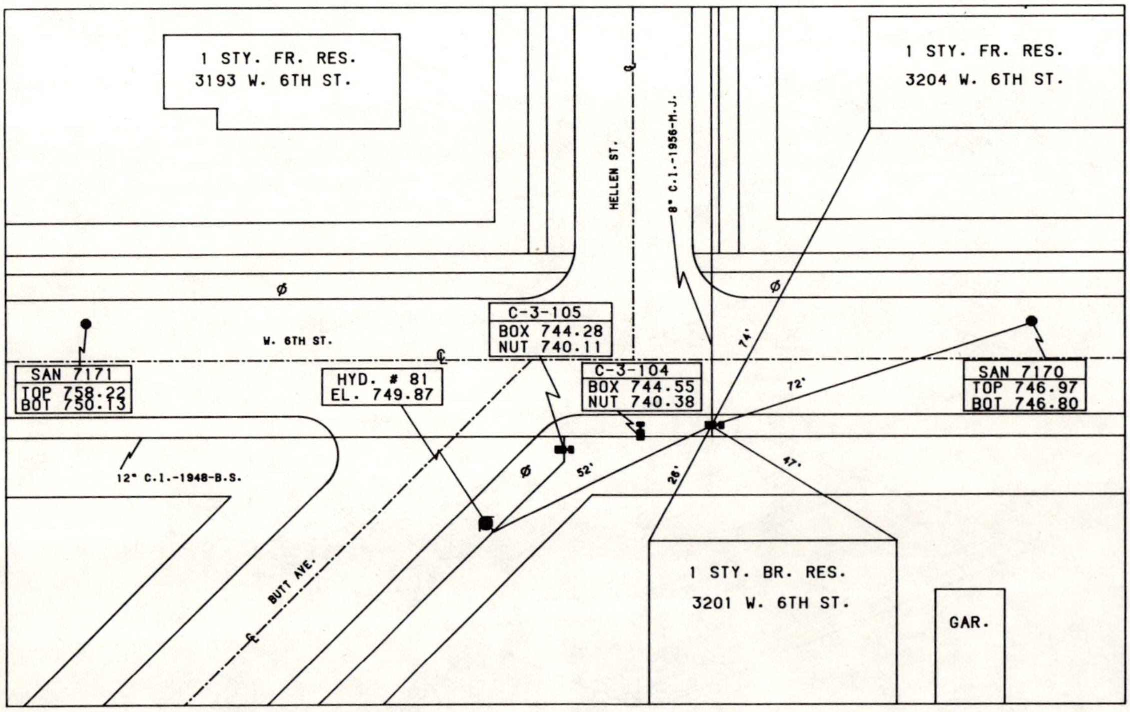

Figure 5-4 Map layer showing valve record.

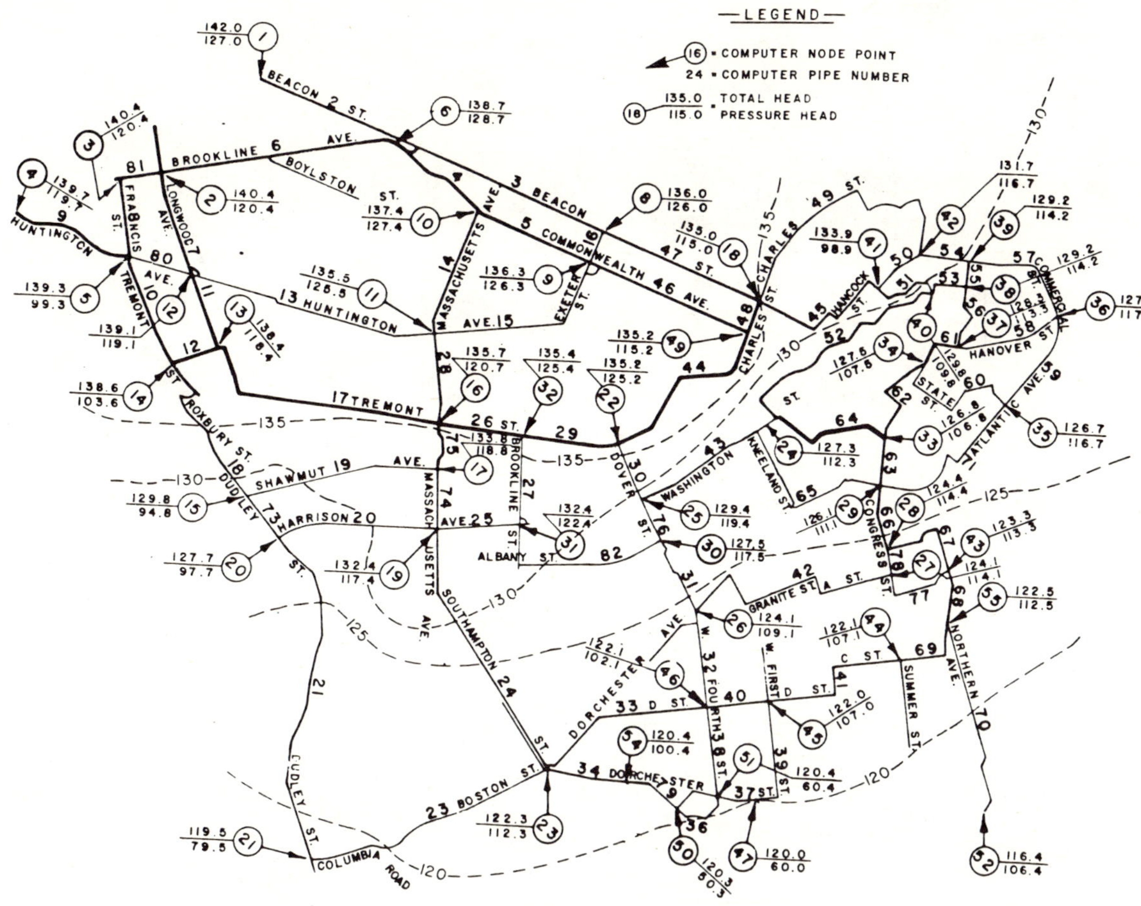

Boston Water Distribution, Report to City of Boston, by Maguire Associates, September 1967. Reprinted with permission.

Figure 5-5 Water distribution system map showing each pipe as line segment.

conditions are more easily interpreted than from analysis programs' tabular output. For systems that include more than one pressure zone, the isolated pressure zones are plotted separately and superimposed on the original plot.

Showing pipes as line segments does not show all of the details of the pipe distribution system. The pipe network may be incorporated with other base maps to include other details, such as smaller-diameter pipes, valves, meters, right-of-ways, streets, and so on.

REQUIRED COMPUTER HARDWARE

Regardless of the specific format used, the graphic presentation of pipe-network data requires special hardware and software. In determining the size and type of computer needed for a particular application, consideration must be given not only to its intended applications but also to the number of terminal hookups, the frequency of use, and the extent of automation of the graphics and other applications. In general, the type of equipment most commonly used in the design, analysis, and operation of water distribution systems include a computer, alphanumeric terminal, graphics terminal, digitizing tablet, plotter, copier, and printer.

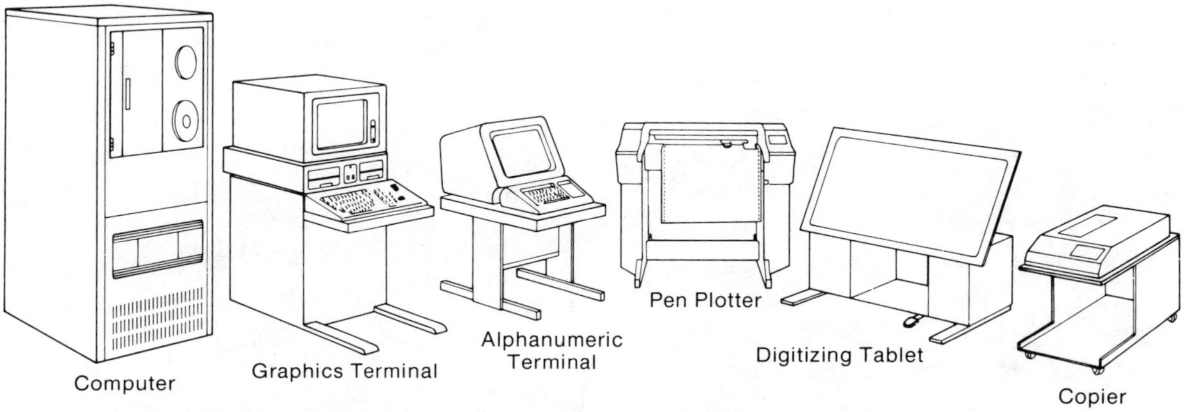

Figure 5-6 Minicomputer configuration.

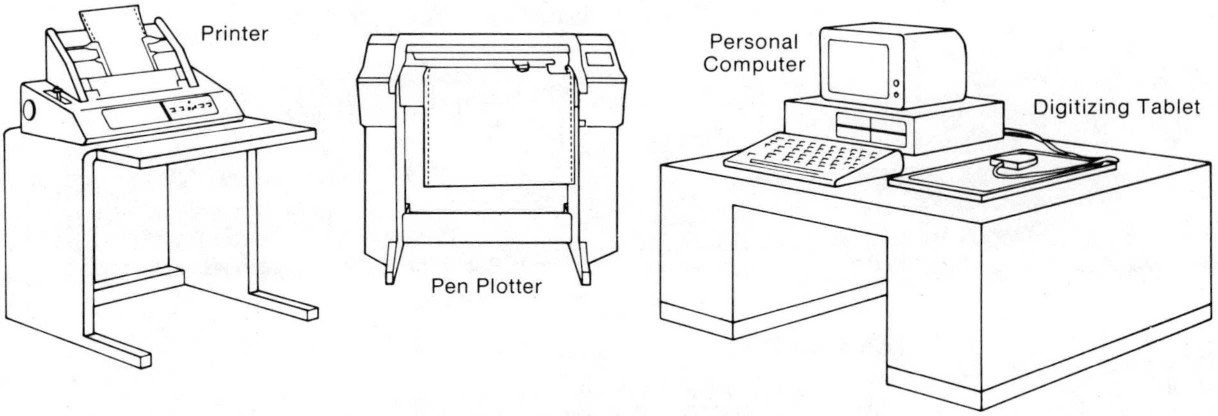

Figure 5-7 Personal computer configuration.

The hardware may be configured for use with a minicomputer or microcomputer (personal computer). With a minicomputer, the graphics terminal, digitizer, and plotter are a separate work station where most of the graphics activity takes place (Figure 5-6). The minicomputer serves as a host computer and is accessible to other users through terminals at other locations. When a personal computer is used to handle analysis and graphics, the graphics and alphanumeric output may be displayed on the terminal or be sent to the printer. A plotter and a digitizing tablet can also be attached to the system (Figure 5-7).

Computer

The computer performs basic machine and housekeeping operations under the control of operating programs. It also processes and manages large volumes of graphics data under the direction of application programs.

Alphanumerics Terminal

The alphanumeric terminal displays alphabetic and numeric characters only. It is used on mainframe and centralized minicomputer systems to key in commands to the computer. It is also used to input data-base information and to modify graphics files produced on the graphics terminal. Graphics terminals can display alphanumeric characters, but this is not their main function. The primary use of an alphanumeric terminal is to manage the computer system. It has no graphics capability.

Graphics Terminal

Graphics terminals fall under two general classifications—vector terminals and raster terminals. Each is used to display graphics information on a video screen by drawing a line between two points defined by X and Y coordinates.

Vector terminals, also known as direct-view storage tubes, have high resolution but lack multicolor capability. Displays on the screen are completely redrawn when any of the graphics data is edited by adding, changing, or deleting a line. Because of this and the advent of inexpensive large memory, vector terminals are becoming obsolete.

Raster terminals have lower resolution but provide multicolor and local editing capability. The graphics information is stored and displayed on the screen as arrays of pixels (or picture elements, pels). Each pixel represents a point on the screen with its X,Y location and color defined by the user through the software. Specific pixels are activated and continuously refreshed in a raster fashion (row by row). The number of pixels is limited by the available memory of the terminal. This limits the resolution of the screen display so that the lines often appear jagged.

Vector and raster terminals may have local intelligence, or memory, that greatly enhances computer operations. Older terminals without local intelligence imposed a heavy demand on the computer and severely limited its use in manipulating graphics data. Today, virtually all graphics terminals have local zoom and pan capabilities to show fine details in a large-scale drawing without having to retransmit the data to the computer.

Digitizers

Digitizers are used to convert graphic information from an existing map or sketch into digital form for entry in the computer. Digitizers in the form of a tablet have a subsurface wire mesh that forms thousands of X,Y coordinates. A map to be digitized is placed on the tablet. A stylus or cursor activates a point on the surface, which generates the numeric value of that point's coordinate (X,Y).

Recently, scanning digitizers have been developed. Scanning digitizers use video technology to capture the lines in a drawing or map all at once. One must still sort through the generated graphics data, however, to select, identify, and label each of the lines in some fashion (that is, water lines, sewer lines, and so on).

Plotters

Plotters produce highly accurate engineering drawings. The quality of the plots depends only on the number of addressable points on the plotter itself. The quality is independent of the resolution of the graphics terminal, unlike the copier. The graphics data generated can be stored and sent, or spooled, to the plotter at a convenient time.

Various types of plotters (drum, belt-driven flat bed, and so on) are available. They can produce drawings in colors using liquid ink or ballpoint pen on vellum or mylar. The width of the plots vary from 20 in. to 50 in., and the length can be as much as 120 ft if a continuous-feed plotter is used. The new electrostatic plotter produces good-quality plots at high speeds.

Copiers

Copiers produce copies of the image exactly as it is displayed on the graphic terminal's screen. Unlike the plotter described above, the quality of the copy depends on the resolution of the graphics terminal and the copier itself. A large variety of copiers produce color or black-and-white copies using electrostatic, ink-jet, and laser-copying processes.

Printers

Printers prepare hard copies of the computer's output and are categorized as being either impact or nonimpact in type. Common impact printers create characters with daisy wheels and dot-matrix devices that strike against inked ribbon and paper. Nonimpact printers create images using thermal, electrostatic, chemical, and ink-jet technology. Using xerography and lasers, nonimpact printers can produce prints at 10 times the speed of impact printers.

REQUIRED SOFTWARE

Software programs fall under two general categories: computer operating systems and application programs.

The computer operating system is a collection of programs designed by the computer manufacturer for the management and overall operation of the computer. Application programs do various tasks specified by the user. For example, in order for the computer to accept graphics data describing a pipe network, a program must convert the signals from the digitizer into numerical values so that the pipe network can be displayed on the graphics screen.

Application programs may be written in various high-level computer languages, but they must be translated into a computer's specific machine code before they can be used. The translation is accomplished through the use of compilers. Graphics-application programs require a special package of programs, known as a graphics library, so that data generated from the application programs can be converted to graphic signals, which are displayed on the terminal screen. The computer operating system permits the user to link applications software with the graphics library during the processing and generation of output. In general, the user need not be concerned with compilation and special libraries. The software supplier takes care of these details by providing the compiled version of the programs.

Graphics Programs

Graphics programs are application programs used interactively. The operation of a typical graphics program used to display pipes and nodes consists of data entry, data editing, and graphics output, or screen display and ink plots.

Data entry. Data is entered using either a digitizing tablet or a scanning digitizer. Pipe-network information is obtained from existing water distribution maps, underground record maps, construction records, or original design drawings. When using a digitizing tablet, the distribution maps (or sketches drawn to scale) are placed on the tablet and the *X,Y* coordinates of each node are registered using a stylus or cursor. The coordinates are tied to a reference system, such as the state plane coordinate or universal transverse mercator systems. Information regarding each node,

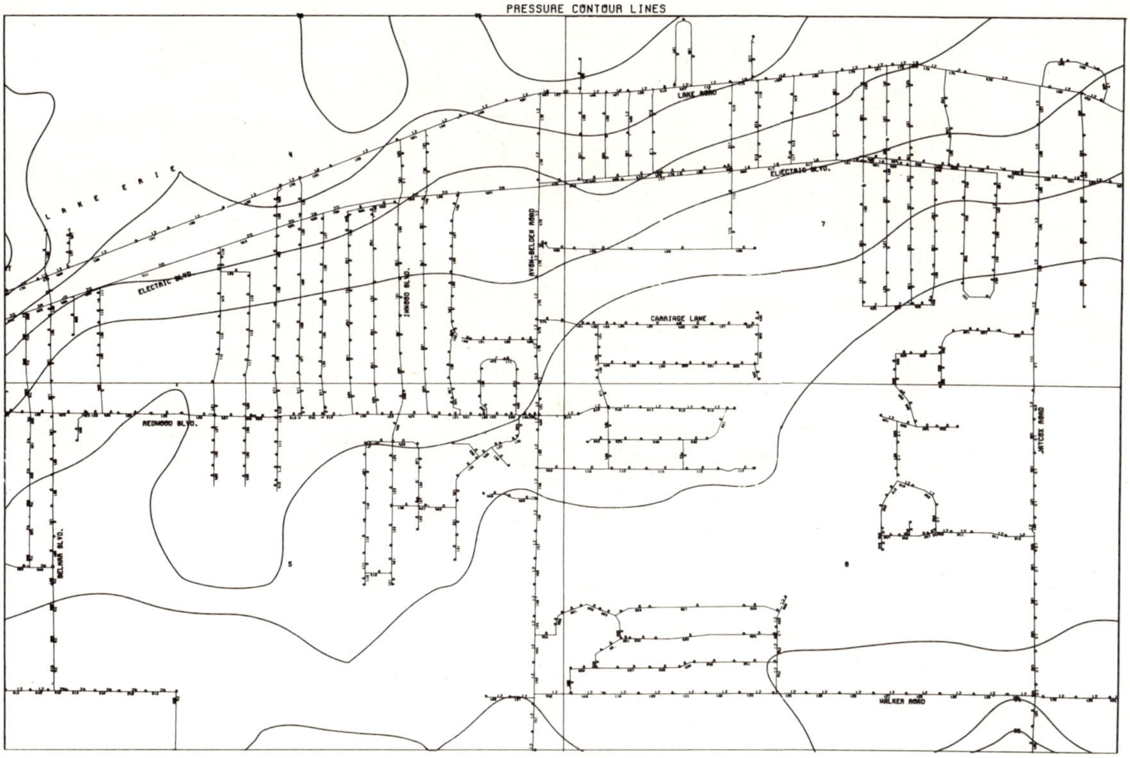

Figure 5-8 Pressure contours obtained from computer simulations.

whether it is a hydrant or a street junction, as well as pipe diameters and pipe/node identification numbers, can then be generated.

Data may also be entered from an electronic field book by unloading it directly to the computer. Another method of entry is through an alphanumeric terminal. Here, the *X,Y* coordinate of each node, based on values obtained from field surveys, is keyed into the computer data file.

Data editing. After it is stored, graphics data can be edited through the graphics terminal or the graphics files themselves. The finished plot may be redrawn on the screen, and each pipe and node may be accessed using a cursor, screen-activated cross hair, light pen, or mouse. The operator can then add, delete, or change the various components of the pipe distribution system.

When editing through an alphanumeric terminal, the files, including numerical values of the coordinates of the digitized points, are brought to the screen to be modified or to have new nodes and pipes added. The updated files are drawn on the graphics terminal to display the changes.

Graphics output. The graphics data files may be viewed on the terminal screen at various levels of detail to compare and verify changes to the system. If color terminals are used, various color patterns may be used to delineate pressure zones, high-demand areas, and so on. The results may also be sent to the copier or pen plotters to make a hard copy. The user may choose various details to be plotted, such as map scales and line thickness, as well as line type and color. Various layers of graphics data, such as right-of-ways and property lines, may also be superimposed on the pipe network plot.

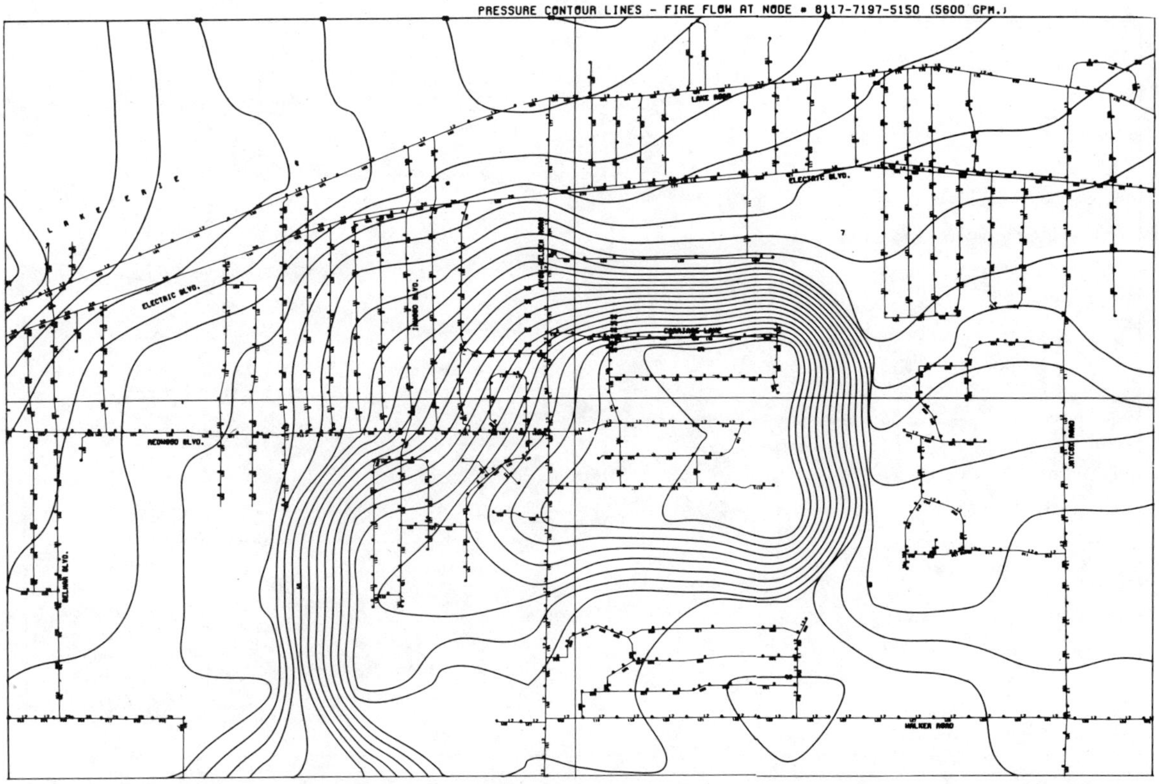

Figure 5-9 Pressure contours obtained from computer fire-flow simulations.

Figure 5-8 shows a pipe network with pressure contour lines for a typical city. Similar contours may be drawn to show the hydraulic grade lines. The contour lines are based on the pressures at each node as computed by the hydraulic-analysis program simulating normal water-usage conditions. Pressure contours may also be generated for special demand conditions, such as a large fire flow at a specified location (Figure 5-9). Such displays give a quick assessment of a system's ability to provide adequate fire flows. They also show flow patterns that are generated while supplying the hydrant, as well as the effect the fire flow has on pressures throughout the system.

Similar fire-flow simulations may be run to determine the capacity of each hydrant in the distribution network. The combined results presented as fire-flow contour lines in gallons per minute (Figure 5-10) can be helpful in assessing the capability of the system under various operating conditions.

Other application programs. Presentation and documentation of model-analysis input and output data is enhanced through the use of word processing, chart, and picture-making software. These application packages run on minicomputers or various standard PCs and printers. Many of these applications are "menu driven" by either a keyboard or mouse. They are also easily backed-up for security and documentation.

Presenting text and illustrations as type, charts, or pictures makes it easy to file and store information on diskettes and microdiskettes and easy to reproduce and duplicate. The format of stored data allows the user to show many alternates in a similar format, allowing for easier comparison and analysis.

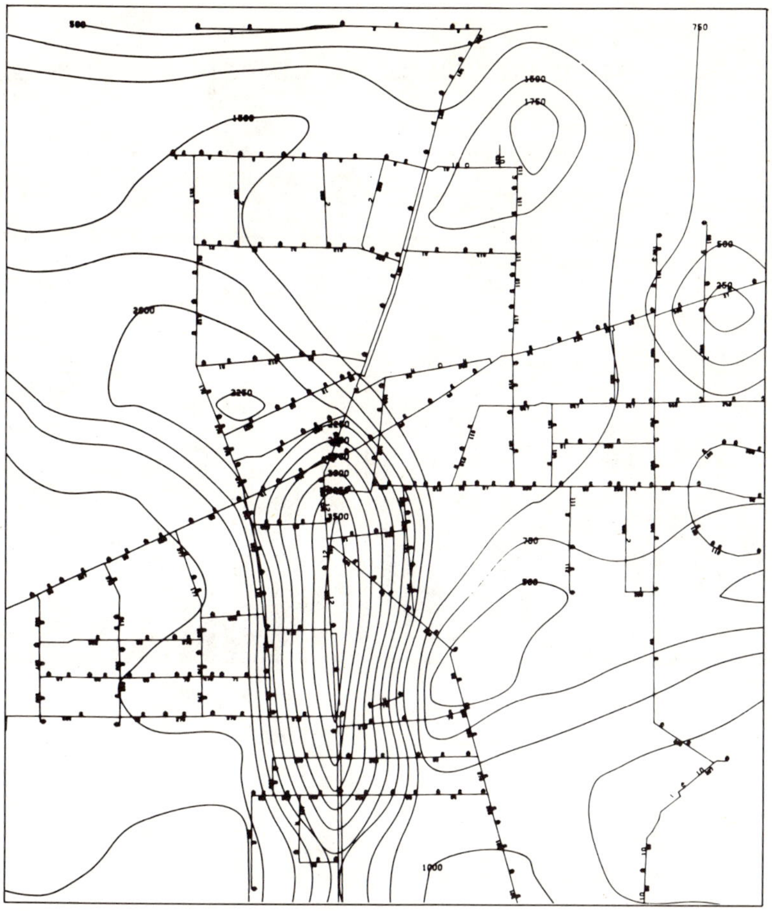

WATER DISTRIBUTION NETWORK - MAP NUMBER 15   FIRE FLOW CONTOUR LINES - GPM.

Figure 5-10 Fire-flow contour line—gpm at 20 psi.

Chart-making software. This type of software translates quantitative data into graphics in the form of bar, line, and column graphs; pie and area charts; or a combination of several graphs and charts. Data is entered into series; each series contains a set of categories, and each category has a corresponding value.

Data is easy to stack or may be superimposed on other graphs for comparisons. This application works well when illustrating tank and demand fluctuations for continuous simulations (see Figures 5-11 and 5-12) and comparing model-simulated results with actual tank levels. Labels, arrows, shading, and user-definable scales and margins allow the user to make high-quality graphs for reports and presentations. Spreadsheet data may also be converted into graphic form with the use of chart-making software.

Illustration software. Minicomputer and PC software for drawing and illustrating defines details of a given component of the water distribution system. The ability to draw sketches, flow-charts, project schedules, and maps in a flexible and easily stored format gives the user a convenient tool for conveying information gathered from the model analyses. Custom-made patterns for shading figures or drawings are also possible.

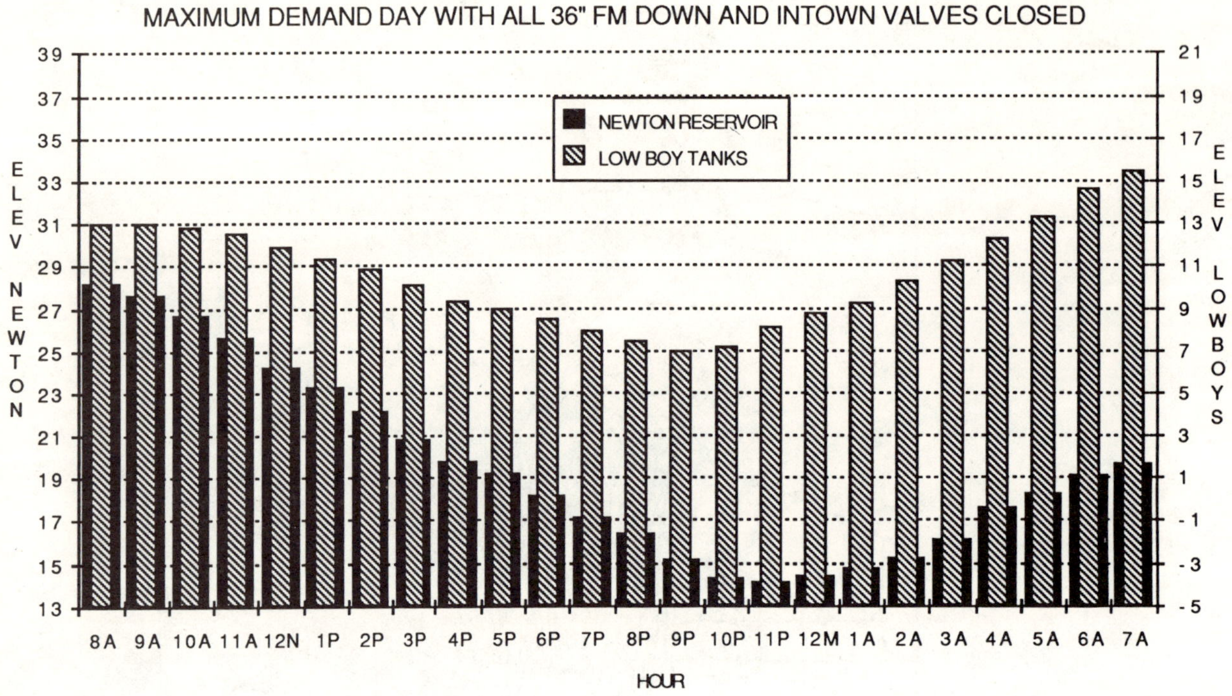

Figure 5-11 Continuous-simulation results.

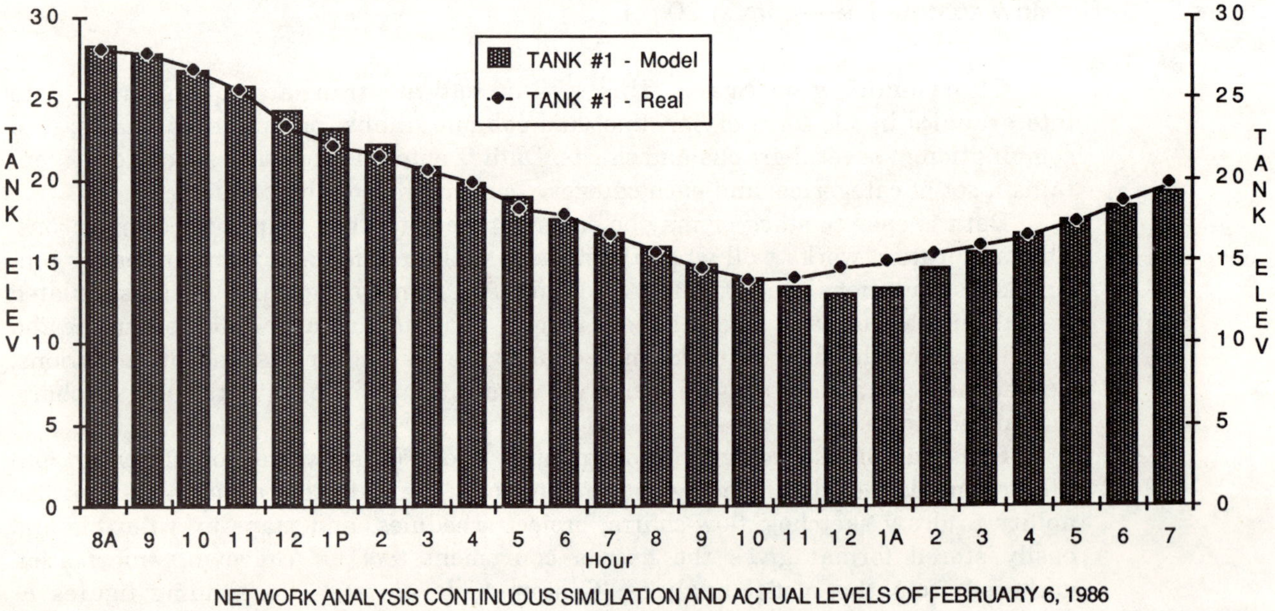

Figure 5-12 Comparison of continuous-simulation results with field observations.

Word Processing Software

Word processing software allows the user to create and store report text, equations, tables, and symbols. Word processing packages allow the user to change and enlarge or reduce fonts, move words and paragraphs, add and delete, and generally edit the text of a report. Text may be made bold or italicized to give it emphasis. Charts and illustrations from software may also be merged into the text. The report format is polished by margin control, tab setting, centering, and line justification. Text or headings that are used often may be stored and recalled from an accessory notepad or scrapbook in the software package.

SUMMARY

Application programs enhance the value of analysis software by displaying its output in a more efficient and comprehensible manner. It also allows the user to create custom-designed professional reports. The quality of computer-graphics software for network analysis varies considerably. The attempt here has been to outline and discuss primary components and does not necessarily represent the best way or the common practice. Nonetheless, the capabilities and procedures addressed herein must be considered when evaluating the use of and need for graphic-support software.

Glossary

applications program Software developed for a specific, practical application. Examples are word processing, data-base management, network solution, mapping, and illustration.

average-day demand Annual water consumption divided by 365 days in a year. The average water demand a given water distribution system experiences over a one-day period. The average-day demand applies to a specific time period.

base data file The data file that contains a comprehensive description of a distribution system's physical and operating characteristics. The base data file is the modeled representation of the existing distribution system.

batch mode A noninteractive operating mode in which the computer executes each command from a file and does not prompt the user for any input until all commands have been completed.

calibration The process of adjusting the input data of a water-distribution-system model to improve agreement between the modeled or predicted pressure and flows and the actual values as observed in the distribution system.

central processing unit (CPU) The "brain" of a computer, which performs the arithmetic and logical operations involved in executing a program.

compiler A computer program that converts the programmer's high-level language into machine code that the computer can execute.

computer mapping The process of using a computer program to draw a map that can be viewed on the terminal's screen or on paper.

data base An organized collection of identically formatted information, generally accessible by computer. The data can be manipulated by the computer under the direction of an applications program.

digitizing tablet or digitizer An electromechanical device resembling a tablet or drawing board that is used to input two-dimensional geometric data, based on an *X–Y* coordinate grid, into the computer.

disk drive An electromagnetic device that magnetically transfers information to and from a diskette or a hard disk.

diskette *See* floppy disk.

diurnal-demand curve A plot of water demand versus time for a 24-h period. Time is plotted on the *X* axis and demand is plotted on the *Y* axis. The curve depicts a typical period of time (average day, maximum day, minimum day, and so on) and is used to simulate the daily operation of the network, especially the cycling of system storage.

equalization storage The quantity of water from storage that enters the system during peak demand periods to augment other sources of supply. Equalization storage is replenished during off-peak demand periods.

extended period simulation (EPS), dynamic state A pipe-network simulation comprising a series of steady-state runs that allow the user to view system operations over a period of time with defined intervals. EPS can model time-varying features, such as tank levels, head or pressure-controlled valves, and status of pumps and pipes.

fire-flow demand The quantity of water required specifically for fire fighting, generally determined by a fire department.

fixed-grade node A node within the model network at which the pressure is specified; also known as a supply node.

floppy disk A thin, flat, circular piece of plastic, coated with magnetizable iron oxide and contained in a square protective cover. The disk is used to store programs and data and is inserted into the disk drive for use.

fragment models A subdivision of a system that can be modeled independently from the larger system. A large, complicated system is sometimes broken down into fragment systems, such as pressure zones, to facilitate analysis.

global adjustments Adjustments made to a particular parameter on a system-wide basis. An example is applying a system-wide percentage change in water consumption to simulate a uniform scaling in demand.

graphics Pictures, diagrams, or graphs, as opposed to the numbers and letters that make up text. Many microcomputers require additional hardware and software to process and display graphics.

hard disk A magnetic mass-storage unit that stores programs and data. A hard disk stores much more data, and makes that data accessible more rapidly than a floppy disk.

hardware The computer's physical components, such as monitors, keyboards, and printers.

hydraulic grade line (HGL) For open channels, reservoirs, and tanks, the HGL is the level of the water surface. For pressure pipes, the HGL is the level obtained by summation of the pressure head and pipe elevation.

Insurance Services Office (ISO) The group responsible for rating fire-protection service districts within cities.

interactive mode A mode of operation in which the software prompts the user for input and acts on the user's response immediately, as opposed to batch operations.

mainframe A large, powerful computer serving many users and capable of performing many simultaneous tasks.

maximum-day demand (MDD) The water consumption, in volume of water, used on the highest consumption day in a year.

maximum-hour demand (MHD) *See* peak-hour demand.

maximum pumping rate (MPR) The maximum pump delivery rate during a specific demand period. The MPR is most significant for systems with limited storage.

maximum storage-replenishment rate analysis An analysis to determine whether the system can adequately replenish storage facilities during low demand periods; also known simply as replenishment analysis.

microcomputer A personal computer, designed for use by one person at a time, which can be placed on a desk.

minicomputer A computer that has more power than a microcomputer but less power than a mainframe.

model The mathematical representation of a distribution system, composed of a network-analysis (or solution) program and a network-data file.

mouse An input device, that is tied to the computer. The mouse is moved about on a flat surface and the cursor on the computer screen follows the movement of the mouse.

network analysis The process of analyzing water distribution systems through the use of mathematical computer models.

networking Connecting computers to allow each to access the other's data bases.

nodes An intersection of two or more pipes where a flow can be added or withdrawn from the system or where the characteristics of a pipe change; a junction.

peak-hour demand (PHD) The maximum total 1-h water demand, in units of volume per day, a given distribution system experienced or would experience during a particular year or other time period. Also known as maximum-hour demand.

plotter A peripheral device that allows a computer to produce hard copies of graphics.

program A set of precise commands telling the computer what steps to perform and in what order.

real-time mode A high-priority mode of computer operation in which the computer program responds to input immediately so that the results can be used to monitor or control physical processes.

sensitivity analysis A procedure for testing the effect of small changes in the assumed values for known heads, consumptions, and pipe or pump characteristics on a model's solutions. Sensitivity analysis can be done automatically by the model or by trial and error.

skeletonization The practice of deleting certain minor pipes and facilities from the system model. Ideally, this creates a manageable model that decreases execution time without losing overall accuracy in model performance.

software Computer programs. *See* program.

source program The version of a computer program in a high-level language. The commands must be converted into machine language before the machine can execute the program. The user can easily modify a source program and list the source program to see how it works.

spreadsheet software A program that formats information (numbers) in a grid of rows and columns. Calculations can be performed and stored within the spreadsheet.

steady-state analysis A pipe network analysis that assumes consumptions, supplies, storage-tank levels, and certain other variables for one specific point in time.

support software Software that can be used to improve, enhance, or otherwise expedite other software.

system analysis *See* network analysis.

transient analysis A pipe network analysis that models pressure waves created by sudden changes in system flow rates.

water audit The process of accounting for the system's consumption by monitoring inflow and outflow measurements at the system boundaries.

Bibliography

ASCE Task Committee on Procedures for Management of Control Systems Engrg. for Water Pipelines. *Management of Engineering of Control Systems for Water Pipelines*. ASCE, New York (1978).

CANENDESE, A. & MELE, P. Optimal Design of Water Distribution Networks. *Jour. Hydraulics Div.—ASCE*, 104:HY2: 237 (1978).

COULBECK, B. & STERLING, M.J.H. Optimal Control of Water Distribution Systems. *Proc. IEE,* 125:1039–1044 (1978).

DAVIS, A.L. & JEPPSON, R.W. Developing a Computer Program for Distribution System Analysis. *Jour. AWWA*, 71:5:236 (1979).

Emergency Planning for Water Utility Management. AWWA Manual M19. AWWA, Denver, Colo. (1984).

EMERY, H.A. ET AL. Introduction to AM/FM-Automated Mapping and Facilities Management. NCGA Proc. 5th Ann. Conf., 351. Anaheim, Calif. (May 1984).

FALLSIDE, R. & PERRY, P.F. Hierarchical Optimization of a Water-Supply Network. *Proc. IEE*, 122:2:202–208 (Feb. 1975).

Fire Suppression Rating Schedule. Insurance Services Office, New York (June 1980).

HOBBS, B.F. Reliability Analysis of Urban Water Supply. In *Computer Applications in Water Resources* (H. Torno, ed.), 1229–1238. ASCE, New York (1985).

JEPPSON, R.W. *Analysis of Flow in Pipe Networks*. Ann Arbor Science Publications, Ann Arbor, Mich. (1976).

KALLY, E. Automatic Planning of the Least-Cost Water Distribution Network. *Water and Water Engineering*, 75:148–152 (1971).

KRAEMEN, K.L & KING, J.L. *Computers in Local Government*. Auerbach Publishers, Pennsauken, N.J. (1980).

LAI, D. & SCHAAKE, J.C. Linear Programming and Dynamic Programming Applied to Water Distribution Network Design. Tech. Rept. 116. MIT Hydrodynamics Laboratory, Boston, Mass. (1969).

NEUMAYER, C.A. JR. Determining Reserve and Equalizing Storages. *Jour. AWWA*, 54:4:394 (1962).

OLEY, J.D. & DAM, A.V. *Fundamentals of Interactive Computer Graphics*. Addison-Wesley, Reading, Mass. (1982).

RYAN, DANIEL L. *Computer Aided Graphics and Design*. Marcel Dekker, New York (1985).

SHAMIR, U. Optimization in Water Distribution Systems Engineering. Mathematical Programming Study, North Holland, no. 11, 65–84. North Holland, New York (1979).

SHAMIR, U. Real-Time Control of Water Supply Systems. Proc. International Symposium on Real-Time Operation of Hydrosystems (T.E. Unny and E.A. McBean, eds.), 550–562. Univ. of Waterloo, Waterloo, Ont. (June 1981).

SHAMIR, U. Computer Applications for Real-Time Operation of Water Distribution Systems. In *Computer Applications in Water Resources* (H. Torno, ed.), 379–390. ASCE, New York (1985).

TOPPING, R.E. & PUCCIA, R.J. A Computer-Assisted Water System Analysis and Design for Charleston, S.C. *Jour. AWWA*, 76:7:83 (1984).

UPMEYER, D.W. Estimating Water Storage Requirements. *Pub. Works*, 109:7:50 (1978).

WALSKI, T.M. Using Water Distribution System Models. *Jour. AWWA*, 75:2:58 (1983).

WALSKI, T.M. *Analysis of Water Distribution Systems*. Van Nostrand Reinhold, New York (1984).

ZESSLER, U. Optimal On-Line Control of Regional Water Supply Systems. Master's thesis, Technion–Israel Institute of Tech., Haifa, Isreal. (1984).

Index